Electro-discharge Technology for Drilling Wells and Concrete Destruction

V. Ya. Ushakov · V. F. Vajov
N. T. Zinoviev

Electro-discharge Technology for Drilling Wells and Concrete Destruction

Springer

V. Ya. Ushakov
Tomsk Polytechnic University
Tomsk, Russia

N. T. Zinoviev
Tomsk Polytechnic University
Tomsk, Russia

V. F. Vajov
Tomsk Polytechnic University
Tomsk, Russia

ISBN 978-3-030-04590-6 ISBN 978-3-030-04591-3 (eBook)
https://doi.org/10.1007/978-3-030-04591-3

Library of Congress Control Number: 2018961723

This Springer imprint is published by the registered company Springer Nature Switzerland AG
The registered company address is: Gewerbestrasse 11, 6330 Cham, Switzerland

Preface

Spark discharge initiated in all dielectric media with electrical voltage exceeding a certain critical value (usually called breakdown) has attracted attention of scientists already since the dawn of the development of electrical power engineering (EPE), and natural discharge (lightning)—since antiquity. At the beginning of the EPE development, the spark was of interest mainly as a cause for the breakdown of insulation of electrical equipment and its failure. With scientific and technical progress in all spheres of material activity and deepening of knowledge about the nature and properties of electrical discharges, the scope of technological application of their manifestation, including the formation of shock waves and radiation in a wide range of wavelengths, high temperatures, chemical and phase transformations, has been expanded.

Since the second half of the twentieth century, a slow, but steady increase in the share of consumption of electricity converted into other energy types by means of electrical discharges of various types in different media has been observed. The high energy intensity of processing and destroying solids and, in particular, superhard non-conducting bodies (of natural and artificial origin) stimulated the birth of the idea of technological application of the spark discharge for destructing and processing of solid non-conducting bodies. The effect was well known to scientists who experimentally studied the electrical breakdown of solid dielectrics (it forced them to replace the samples after each breakdown) as well as experts in high-voltage engineering forced to replace or restore equipment after insulation breakdown.

The birth of the idea of the application of the spark discharge channel in a solid at Tomsk Polytechnic University (TPU) was natural, since here the scientific school on the study of nature and main regularities of electrical breakdown of solid dielectrics was founded under the leadership of Prof. A. A. Vorob'ev in the 1950s. These studies were performed at the Department of High-Voltage Technology (DHVT) founded in 1946. The number of investigators and the scientific scope of these studies expanded as results of investigations of physics of solid dielectric breakdown embodied in the development of a new electro-discharge technology (EDT). By 1963, achievements in this field had provided the basis for establishing the "Kedr" Laboratory. Employees of a number of other TPU divisions, whose scientific potential contributed to a solution of the complex problem of EDT development, were actively involved. For better coordination of scientific research, the High-Voltage Research Institute (HVRI) was founded in 1968 based on the "Kedr" Laboratory and TPU divisions of high-voltage technology, electrical insulating and cable technology, and technology of exploration of mineral deposits.

Already, the very first results of testing of installations implementing the new technology under laboratory and field conditions were so impressive, especially when drilling wells, that they inspired confidence in rapid and large-scale EDT introduction in practice. For example, when drilling wells 150.0–180.0 mm in diameter in 1968, penetration rates of 6.05 m/h were achieved at energy consumption of 1.57 kW h/m. The possibility of drilling deep wells (up to 233 m) with diameters up to 235 mm was demonstrated.

In the granulation of minerals and artificial materials, such EDT advantages were revealed:

- selectivity of destruction—preservation of minerals from destruction and more complete extraction of useful components,
- absence of contamination of the final product by materials of working tool destruction,
- possibility to regulate the particle size distribution of the product, i.e., to obtain a narrower particle size distribution than in mechanical granulation.

In breaking and cutting of rocks and concretes and destructing reinforced concrete products, the EDT advantages include:

- utilization of substandard reinforced concrete products due to the possibility of separating concrete from reinforcement without deformation that allows reusing of both concrete and reinforcement,
- possibility of obtaining a curved cut trajectory when moving the electrode system.

With all these EDT applications, its advantages are:

- high energy efficiency of technological processes due to a smaller number of supplied electric energy conversion into the destruction energy,
- weak dependence of the technical and economic indicators of crushing and grinding of rocks and artificial materials on the hardness and abrasiveness of the material or product being processed/destroyed,
- high wear resistance and safety of the working tool (electrode system).

However, the reality turned out to be different—after a successful start, the progress in the achieved results slowed down, which nearly destroyed this unique technology. The reasons for this will be analyzed in Chap. 1. Here, we only note that due to insignificant progress in the most attractive scope of EDT application—in drilling wells—the HVRI began to reorient itself to the development of other directions of EDT application. The scope of work on ED drilling had been greatly reduced.

It took time to rethink the strategy of EDT application in drilling operations, as well as of wider cooperation of partners, including foreign ones. It became clear that success could be achieved only by focusing efforts on the most promising areas of EDT application in drilling and using the potential of international cooperation to solve this difficult complex problem. Until the early 1990s, it had been impossible due to secrecy of the subject that excluded the publication of materials in the public press and their discussion not only with foreign, but also with Russian specialists; the notorious "Iron Curtain" supported the isolation regime. Only since the early 1990s, it has become possible to inform foreign scientists about the development of a new electric discharge technology in the USSR [1–4] and to give objective analysis of problems encountered by the developers of this technology [5, 6]. Cooperation was proposed also for solving problems that hinder wide practical application of the new technology.

A new stage in the EDT history goes back to 1996 when contracts on testing the technology of destruction of substandard reinforced concrete products and drilling wells with a diameter of at least 300 mm in particularly strong rocks were concluded with Kobe Steel, Ltd and Komatsu, Ltd. In the process of fulfilling the contract, the Japanese party organized patenting of inventions of employees of the HVRI in other countries. Three international applications for inventions were submitted, and two patents were received in the USA, one in Australia, and three in Japan.

By the time of completion of collaboration with Japanese firms, the HVRI had acquired new customers—Statoil ASA and UnoDrill, Ltd (Norway), Schlumberger (USA, England, and France), Technical University of Dresden, and Swiss Geo Power Engineering AG. High requirements of the customers (on the diameters and depths of wells, efficiency of destruction of concrete products, and reliability of equipment) determined the need for further deepening and expanding scientific research and design studies. Their results are presented in this monograph together with achievements of preceding decades.

Since practically all materials on the EDT obtained at Tomsk Polytechnic University and partner institutions (Karaganda Polytechnic Institute, Kola Scientific Center of the Russian Academy of Sciences, and the Mekhanobr Institute) for several decades were presented in confidential sources (dissertations, scientific and technical reports, and hard-to-reach conference proceedings), it was necessary to generalize and to systematize these materials in monographs accessible to a wide range of readers. The opportunities appeared in modern times were realized in publication of three monographs [7–9]. They were devoted to consideration of physical foundations of EDT, transient processes in charging and discharging circuits of technological high-voltage impulse installations, and technical and technological implementation of basic EDT ideas primarily for crushing and disintegration of solids. Technologies of well drilling, destroying reinforced concrete products for utilization of their components are presented in them, mainly to illustrate the possibilities of EDT. In this monograph, attention is focused on the results of field tests and partly on laboratory experiments with actual EDT objects—

rocks, concretes, solid insulation, and washing liquids. A significant part of them has been obtained after publication of the above-mentioned monographs.

The main objective persuaded by the authors of new monograph [10] devoted to the EDT was to present in a systematic way the materials on the EDT for wells drilling, destruction and recycling substandard reinforced concrete products, cutting and breaking of rocks and concretes and to attract attention of managers of industrial enterprises and firms to the new technology in the hope of its real support and promotion in industry. For the same purpose, the authors decided to publish its English version in order to inform a wider range of experts and scientists about the new technology. The limited availability of monographs [7–9] for a foreign reader compelled us to repeat (by permission of the authors) in [10] and its English-language version some materials presented in them concerning first of all physical principles of EDT.

We consider it our duty to remind that the founders of the EDT are Profs. A. A. Vorob'ev, G. A. Vorob'ev, A. T. Chepikov, and immediate heads of teams—developers of the technology are I. I Kalyatsky, S. S. Sulakshin, Yu. B. Fortes, N. E. Kovalenko, B. V. Siomkin, V. I. Kuretz, S. Ya. Ryabchikov, and N. F. Pobezhimov. During intensive work on EDT, one of the authors of this monograph (V. Ya. Ushakov) was Director of the HVRI, the other two authors (V. F. Vajov and N. T. Zinoviev)—Heads of Laboratories (their brief biographies are presented on the back cover).

Continuation of works on EDT in the last two decades became possible thanks to the active support of the Directors of the HVRI (now its name is School of Advanced Manufacturing Technologies of Tomsk Polytechnic University) A. A. Dulzon, S. G. Boev, V. V. Lopatin, A. N. Yakovlev, and Head of the Laboratory V. M. Muratov.

The monograph was translated by Ph.D. L. G. Shamanaeva. In the preparation of the manuscript for publication, great help was rendered by engineer E. V. Bogdanova. Preparation of the monograph for publication became possible due to financial support of the Rector of TPU, Prof. P. S. Chubic. The authors express their heartfelt gratitude to each of them.

Readers can send their comments and suggestions to e-mail address: vyush@tpu.ru. They will be accepted with gratitude.

Yours sincerely

Tomsk, Russia

V. Ya. Ushakov
V. F. Vajov
N. T. Zinoviev

References

1. Boev S, Vajov V, Jgun D et al (1997) Electropulse technology of material distraction and boring. In: IEEE international pulsed power conference, USA, Baltimore, pp 220–225
2. Boev S, Vajov V, Jgun D et al (1999) Destruction of granite and concrete in water with pulse electric discharges. In: 12th IEEE international pulsed power conference, Monterey, USA, vol 2, pp 1369–1371
3. Boev S, Vajov V, Jgun D et al (2000) Research of conditions of material destruction by electric pulse discharge. In: 4th Korea–Russia international symposium of science and technology, part 3, (Korus 2000), Ulsan, Korea, pp 346–349
4. Dulzon A, Vajov V, Jgun D et al (2001) Electropulse wells drilling. In: International conference on pulsed power applications. Gelsenkirchen, Band 2, p E06/1–E06/5
5. Dulzon AA, Ushakov VY (2001) State and development perspective of technological applications of high voltage high-powered pulse. In: International conference on pulsed power applications, Gelsenkirchen, 27–29 March 2001 (paper number A. 02)
6. Ushakov VY, Dulzon AA (2001) Performance capability of technological installations using the electrical discharge or conductor electrical explosion energy. In: International conference on pulsed power applications, Gelsenkirchen, 27–29 March 2001 (paper number A. 03)
7. Siomkin BV, Usov AF, Kuretz VI (1995) Bases of electropulse destruction of materials. Nauka, St. Petersburg, 276 p (in Russian)
8. Usov AF, Siomkin BV, Zinoviev NT (2000) Transient processes in installations of electropulse technology. Nauka, St. Petersburg, 160 p (in Russian)
9. Kuretz VI, Usov AF, Zuckerman VA (2002) Electropulse disintegration of materials. Apatity: Publishing House of the Kola Scientific Center of the Russian Academy of Sciences, 324 p (in Russian)
10. Vajov VF, Zinoviev NT, Ushakov VY (2016) In: Ushakov VY (ed) Electrodischarge technology for drilling wells and fracture of reinforced concrete products: monograph. Publishing house of Tomsk Polytechnic University, 312 p (in Russian)

Contents

Abstract

The monograph generalizes results of many-year research and development of perspective electro-discharge technology with application to drilling of wells, destruction of reinforced concrete products, and cutting of cracks in rocks and concrete. Electro-physical bases of this technology and technical implementation of the idea of application of a spark discharge as a "working tool" in rocks and concrete for realization of the above operations are considered.

The monograph is addressed to scientists and experts working in the field of search, exploration, and extraction of resources and building of new and reconstruction of old building objects. It can be used as a textbook by students and post-graduates for deepening knowledge in the field of innovative technologies.

Chapter 1
Relevance, Physical Basics, and Problems of Large-Scale Introduction of Electrodischarge Technology

1.1 Destruction of Solid Dielectric Materials by Traditional Methods

In this Chapter it is expedient to use the collective term *destruction of solid dielectric materials* understood as a number of technological operations using an electric discharge as a *working tool* for drilling wells, destruction of reinforced concrete products, granulating of the filler extracted from concretes in the process of their recycling, cutting, and breaking of rocks and concretes.

As is well known, rocks and concretes form the following series by the order of decreasing resistance to mechanical treatment of different types: compression—bending—abrasion—tension. The tensile strength of rocks is approximately 10 times less than the compression strength.

All traditional destruction methods have the common disadvantage—static or shock loading is applied to the body being destructed with prevalence of compression or cutting stress. In this regard, high pressure of the working tool on the material is required for its destruction; as a consequence, it has low resistance, fast abrasive wear, and high specific energy consumption.

The main requirements imposed on new methods of material destruction are the following:

(1) solids should be destructed without strong mechanical (shock or friction) impact of the tool on them (restrictions imposed by their resistance and high power consumption for friction should be removed),
(2) solids should be destructed in the pulsed mode that eliminates considerable power consumption on plastic deformation and allows high power to be concentrated with low initial amount of energy,
(3) adjustable degree of granulation eliminating energy consumption for regranulation of destruction products should be ensured,
(4) solids should be loaded with prevalence of tensile and shear stresses over compression ones,

V. Y. Ushakov et al., *Electro-discharge Technology for Drilling Wells and Concrete Destruction*, https://doi.org/10.1007/978-3-030-04591-3_1

(5) intermediate energy conversion accompanied by its considerable losses should be eliminated,
(6) good controllability over the character and direction of destruction should be provided,
(7) high drilling speed and low energy consumption should be ensured,
(8) wide range of well diameters and shapes should be covered,
(9) low wear of a drilling head should be guaranteed,
(10) potential capability of drilling wells with depths up to thousands of meters must be provided (today the guaranteed depth is 3500 m),
(11) capability of drilling with a core must be ensured,
(12) a drilling rig should be mobile,
(13) wide capabilities for regulation and automation of drilling process should be provided.

Electrodischarge technology meets most of the above-listed requirements.

During ***drilling*** for well penetration, especially high requirements are imposed on non-failure operation of destroying tool (drilling crown) due to high financial expenses and great loss of time because of its high cost and cumbersome procedure of its replacement in case of failure. Drilling of rocks is one of the most expensive and labor-consuming technological processes in exploration and mining of mineral deposits.

Technical and economic analysis of modern methods of well drilling demonstrates impossibility of elimination of a number of their disadvantages [1–5] causing high cost of drilling, including:

- Low efficiency of 2–10%,
- Complex kinematic systems for transport of mechanical rotation energy to the well site,
- Drilling head that wears out quickly and is insufficiently reliable,
- A sharp decrease in the drilling rate with increase in the hardness of rocks being destructed and in the well diameter, Table 1.1 [6].

During ***destruction*** of reinforced concrete products, ***cutting***, and ***breaking*** of solid nonconducting bodies, the main problems are the same as during drilling, but with their own specificity. Traditional mechanical methods are based on four main destruction types: granulating, splitting, abrasion, and impact, Fig. 1.1. The forces required to decompose the material are produced by granulating machines whose design provides destruction of a certain type.

However, in modern practice of product dispersion, abrasion and bending are used in granulating machines from design considerations in addition to granulation and shock. Exactly due to these reasons, the granulation and disintegration operations, based on the mechanical method, are energy-intensive and costly for reworking on mineral enrichment plants. The relative contribution of these operations to the total cycle of material treatment reaches almost 60% [6, 7]. In this case, the electric energy consumption for material granulation reaches approximately 10–20 kW h/t.

Table 1.1 Energy capacity of well drilling by different methods

Drilling method	Specific energy of destruction, J/cm^3 (kg/cm^3)
Drilling with special rock-cutting tools	
Shock	200–650 (20.4–66.3)
Rotational:	
– Drilling with cutters	700–950 (71.4–96.8)
– Diamond drilling	600–800 (61.2–81.5)
– Shock-rotational	400–600 (40.8–61.2)
– Rotational-shock	600–800 (61.2–81.5)
Drilling without special rock-cutting tools	
Explosive	200–400 (20.4–40.8)
Hydraulic[a]	1000–2000 (102–204)
Hydro-pulse[a]	70–100 (7.1–10.2)
Electro-hydraulic	400–500 (40.8–51.0)
Electro-thermal[a]	5000 (510)
Electro-pulse[a]	100–200 (10.2–20.4)
Laser[a]	5000–12,000 (510–224)
Combined drilling	
Fire	1500 (153)
Plasma	5000 (510)
Thermo-shell	1200–1600 (122–163)
Electro-thermomechanical	500–800 (51–82)

[a]Methods intended for borehole and large-diameter well penetration

One more essential disadvantage of mechanical destruction is nonselective destruction that requires finer grinding to extract the useful component and, as a consequence, leads to increased energy consumption.

By the present time, more than 40 methods of destructing rocks, concretes, and other mechanically resistant materials have been developed. Despite considerable progress in this area, the efficiency of traditional methods of destruction remains low with high cost of works and structural materials.

1.2 Physical Basics of EDT

Breakdown of solid dielectrics, just like other dielectric materials, is terminated by interelectrode flashover of the plasma channel in which the electric energy stored in the source of high-voltage pulses is liberated for 10^{-6}–10^{-5} s. In the process of energy liberation, the channel expands creating the mechanical stress field changing with

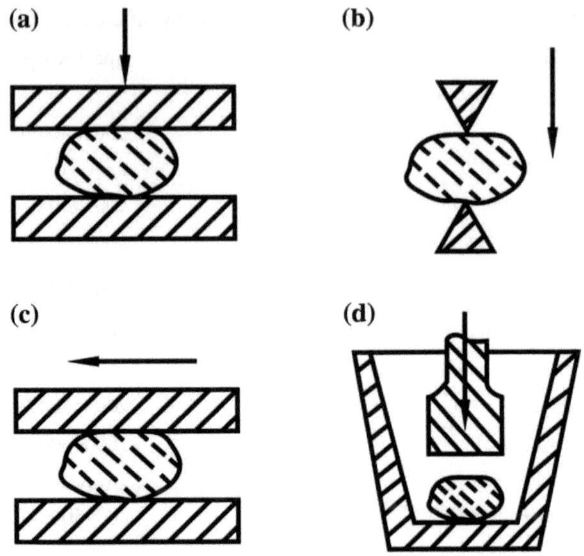

Fig. 1.1 Schemes of mechanical methods of solid destruction by granulating (**a**), splitting (**b**), abrasion (**c**), and impact (**d**)

time and space. This field has a complex character with large relative contribution of tensile and shear stresses in case of breakdown of solids due to intensive divergence of waves and their interaction with free boundaries and inhomogeneities of the solid. Processes analogous to explosion develop in the bulk of the solid. (To describe the physical bases of electrohydraulic and electrodischarge technology, the discharge channel in a liquid or solid is often considered by analogy with an explosive charge [4, 8, 9]). The *working tool*—the plasma of the discharge channel—is being renewed from discharge to discharge, i.e., *does not wear out or aged*. This method provides the possibility of achieving higher destruction efficiency and lower energy consumption in comparison with the mechanical methods. By analogy with explosive destruction, the stress source in EDT is placed directly into the bulk of the solid being destructed with prevalence of tensile stresses. The dynamic character of loading provides brittle destruction of the material with very low energy losses on plastic deformation.

The EDT is implemented primarily by creating conditions for the discharge channel penetration into the bulk of a solid dielectric and its subsequent development inside of the dielectric rather than on its surface or in the surrounding [10]. The authors of the idea of spark application as a working tool [10] drew attention to the effect of a higher rate of increase of the liquid dielectric strength compared to the solid dielectric strength with decreasing time of pulsed voltage application. This leads to the fact that upon exposure to voltage pulses whose front is shorter than 10^{-6} s, the electric strength of dielectric liquids and even technical water is greater than that of solid dielectrics and rocks.

Figure 1.2 shows the voltage time characteristics (VTC) of solids and liquids and the voltage pulses ($U(t)$) generated by the source and recorded in the process of breakdown of the interelectrode gap $U_{br}(t)$.

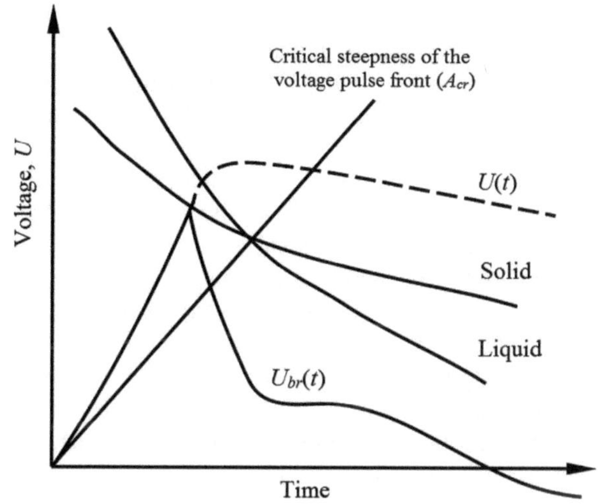

Fig. 1.2 Illustration of the main idea of electrodischarge technology of nonconductive solid destruction

The actual (experimental) voltage time characteristics of some rocks and liquids are shown in Fig. 1.3.

Point A_c of the VTC intersection corresponds to equal strength and probability of electric breakdown of the material being compared. When the voltage pulse exposure time is shorter than 10^{-6} s, the rock becomes electrically weaker than the transformer oil, and when the exposure time is shorter than (2–3) 10^{-7} s, it is weaker than technical

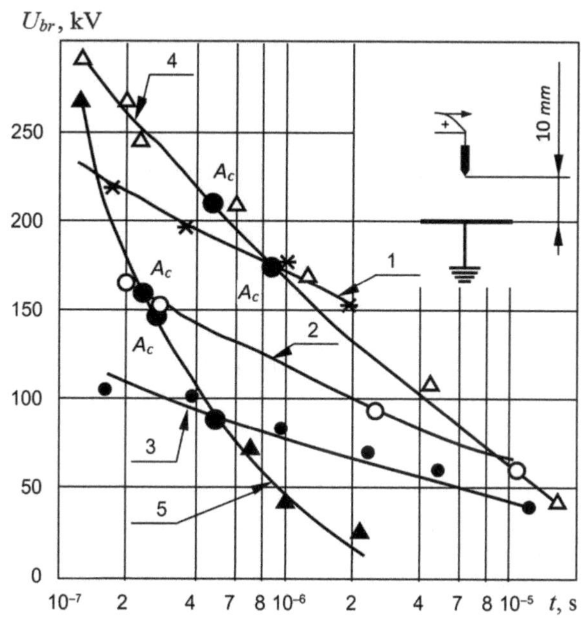

Fig. 1.3 Voltage time characteristic of some rocks and liquids. Here 1 is for quartz, 2 is for felsite-porphyry, 3 is for transformer oil, 4 is for marble, 5 is for clay slate, 6 is for sandstone, and 7 is for water at $\rho = 6 \times 10^3$ Ω cm

water. To the left of the point A_c in the parallel combination of rocks and liquids the solid rather than liquid breakdown is most probable. The slope of the voltage pulse front corresponding to the VTC intersection point (A_c), that is, to the condition of equal strengths of the materials, is the main criterion for application of the EDT technology.

In liquids with high electric conductivity and dielectric constant, the condition for discharge penetration into a solid is less favorable—to provide penetration of the discharge channel into the solid, voltage pulses of higher amplitude and shorter front are required (for example, for technical/tap water, pulses with 10 times shorter front (10^{-7} s) are required compared to transformer or diesel oil. This is due to the fact that for the source of voltage pulses, the working tool in liquid with high electric conductivity is a low ohmic load deforming the voltage pulse and acting directly on the solid being destructed/treated (this reduces the amplitude and slope of the pulse front). High dielectric constant causes the electric field distribution unfavorable for penetration of the discharge channel.

The position of the discharge channel, electrodes, and solid being destructed in different EDT schemes is shown in Fig. 1.4. By the present time, the breakdown mechanism and the basic laws of its development in solid and liquid dielectrics have thoroughly been studied. However, there is still no convincing explanation of the mechanism and observed regularities of the discharge channel penetration into a solid dielectric in an insulating liquid, though the corresponding attempts were undertaken [6]. The efficiency of electro-discharge destruction of solids is determined primarily by the probability of channel penetration into the solid and by the specific energy consumed for destruction.

The electric strengths of rocks differ not so significantly as their mechanical strengths, and it is quite reasonable to suggest a weak dependence of ED destruction characteristics on the rock strengths. Consequently, the EDT is especially advantageous in comparison with traditional mechanical technologies when the body being destructed has enhanced strength.

Figure 1.5 shows the diagram illustrating changes of the ED drilling speed of rocks by uniaxial compression versus their strength. It can be seen that when the compressive strength of rocks increased by 17 times, the ED drilling speed decreased only by a factor of 2.6, whereas for mechanical drilling, such increase of the rock strength caused the drilling speed to decrease by 12–15 times.

Practically all rocks and solid dielectric materials (including elastic ones after their embrittlement by deep cooling), except materials with a very high content of electroconductive (metal) inclusions can be destructed by the ED method. The electric energy of voltage pulse generators (VPGs) is converted into mechanical destruction work directly in the rock without intermediate stages, and the rock is cleaved.

To implement the ED destruction technology, the high-voltage pulse source with proper time and energy characteristics is required. Schemes of forming high-voltage pulses for EDT should meet two requirements: in the stage of breakdown initiation and development, the guaranteed generation of voltage pulses with preset amplitude and pulse slope is required, and in the final stage (the stage of source discharge through

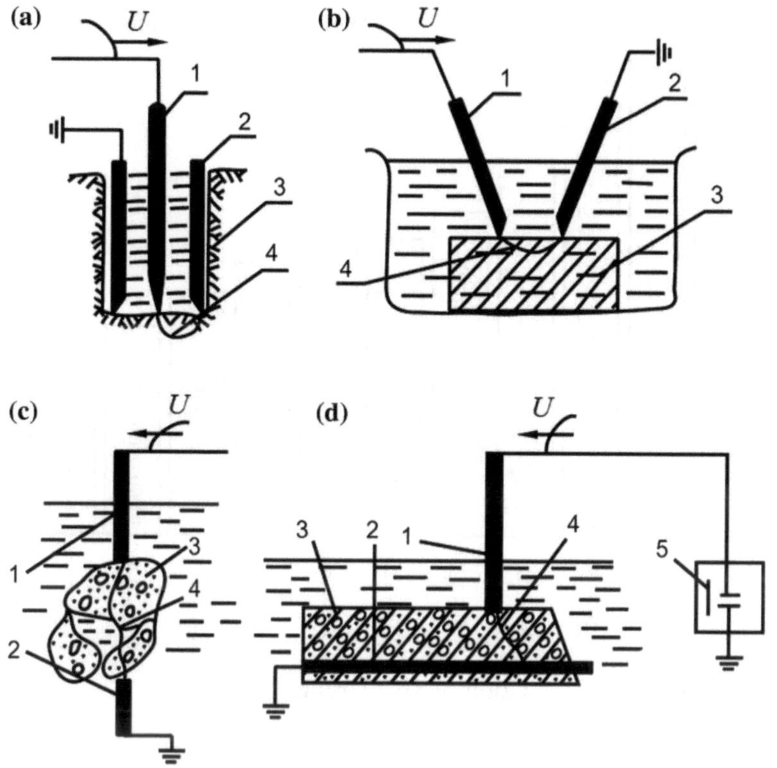

Fig. 1.4 Position of electrodes and destructible solid and discharge channel trajectory for EDT of well drilling (**a**), cutting (**b**), granulation and disintegration (**c**), and destruction of reinforced concrete products (**d**). Here 1 is the potential (high-voltage) electrode, 2 is the earthed electrode, 3 is the destructible solid, 4 is the spark (discharge) channel, and 5 is the high-voltage pulse source

the plasma channel), provision of the optimal regime of energy deposition into the channel is required. During laboratory investigations, different schemes of high-voltage pulse generation (capacitive multistage voltage pulse generators—VPGs, sources with inductive energy storage, and sources with cable forming lines) and different pulse shapes (oblique bell-type pulses, lightning-type pulses with oblique front and long pulse decay time, and rectangular pulses with nanosecond front and slow pulse tail voltage decay) were approved.

In EDT, capacitive multistage VPGs (Marx-generators) that differ by simple design and service and allow voltage pulses with amplitude of 200–600 kV and energy per pulse of $0.1–50 \times 10^3$ J are most widely used. They had no alternative at the time of EDT origin and still do not have a worthy alternative, in particular, for EDT applications considered in the monograph.

For drilling, the VPGs mounted on the surface rather than inside of the well were mainly used. The development of pulse sources with such voltage level that are immersed into wells represents an independent challenge. Thereupon, the drilling

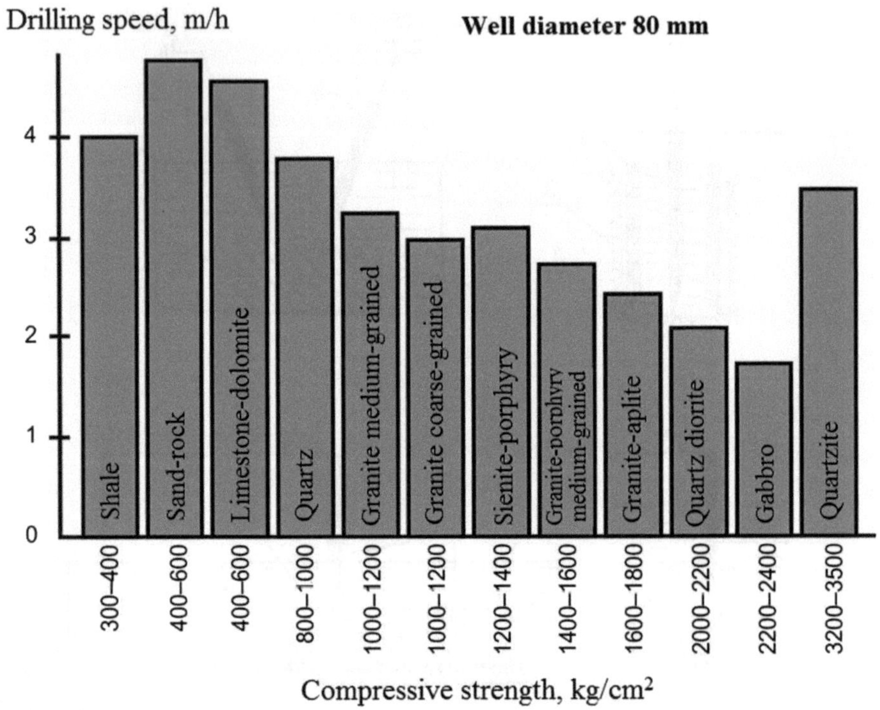

Fig. 1.5 Diagram of change of the ED drilling speed versus the compressive strength of rocks (for a well diameter of 80 mm)

rig should perform not only its characteristic functions, but also be a transport line between the VPG and the drilling head. This naturally creates additional problems, first of all, connected with electric strength of insulation and energy losses. The actual efficiency of energy transport from the energy store is in the range (0.35–0.75) [6, 9]. Here by the efficiency we understand the ratio of the energy liberated in the discharge channel to the energy stored in the VPG.

Among the disadvantages of this scheme are also limited possibilities for changing the regime of energy deposition and a deeply oscillating regime of the discharging circuit. Some other serious disadvantages of the VPG as a power supply for the EDT are considered below.

As follows from the EDT operating principle, ED drilling needs no rotation of the drilling head and other mechanical loads; wells of arbitrary shapes (round, square, elliptic, etc.) can be drilled; the well area has practically no restrictions; and core-drilling (geology prospecting drilling) is possible. Schemes of flushing out wells and of slug removal from wells differ insignificantly from the conventional schemes used for mechanical drilling. Drilling head wear is insignificant, which allows wells several hundred meter deep to be drilled without its replacement; in addition, it can be made of ordinary steels (for more detail, see Chap. 7).

Based on the accumulated experience and recently obtained results, we can assert that the ED ***drilling*** method completely meets the requirements formulated above.

The EDT ***of destructing sub-standard reinforced concrete products (RCP)*** has a number of significant advantages over the mechanical technologies:

- Metal framework that serves one of the electrodes does not undergo limiting deformations and preserves its form;
- Technological process takes place under the layer of water, which excludes dust formation;
- Mass and size of the ED installations exceed by 2–3 times those of mechanical RCP destruction installations;
- Full recycling of destruction products is possible, thereby yielding positive economic and ecological effects.

The ED technology ***of cutting and breaking*** of rocks also has a number of advantages:

- Low energy consumption;
- Capability of figured cutting of stone and artificial materials;
- Stripping of rock bases for buildings when constructing new facilities, especially in cities, and for tunneling with restrictions on blasting;
- Capability of direct cutting of rock blocks from a rock massif without explosive breaking;
- Capability of penetrating cracks, trenches, and pits.

1.3 Problems of Large-Scale Implementation of EDT

The first works on ED drilling of wells in rocks of laboratory samples (1961–1963) and then in the vicinities of the cities of Tomsk, Apatity, and Leninogorsk, on Kolyma, on the Baikal–Amur Railway route, and around Kursk magnetic anomaly (1963–1978) demonstrated that the new technology meets most requirements formulated above. During field drilling tests, more than 1000 m of wells were penetrated (maximal depth of one well was 233 m) and the rate of penetration equal to 14 km/h (in permafrost on Kolyma) was attained when drilling the well with a diameter of 250 mm.

Test operation of the drilling unit (working mockup) based on new principles allowed its advantage by many indicators to be proved, but simultaneously revealed a number of problems connected with its finishing to semi-industrial standards and especially, its mass production.

To analyze the reasons for slow EDT advancement in production process, it is expedient to combine them, in a certain measure tentatively, into two groups—*objective* and *subjective* [8, 11].

Among the *objective* reasons are:

1. The USSR industry in the second half of the XXth century was little sensitive to innovations in civil industries under conditions of planned economy (with characteristic absence of actual competition inside the country) and Cold War (spurring on the arms race). In this regard, final stages of their development in civil engineering—design of installations and organization of their industrial production—faced almost insurmountable obstacle. In this connection, it should be considered that EDT installations are complicated and large-sized technological complexes (in particular, drilling rigs).
2. High power, high voltage pulse technology at that time did not fully meet the requirements imposed by EDT (and in many respects, it does not fully meet these requirements even now). First of all, it is a question of a compulsory long (many million pulses) operational lifetime and high pulse repetition frequency (no less than 10 pulses/s) of pulsed high voltage sources. Because of lack of other sources, multistage high-voltage pulse generators (VPGs), initially intended for testing insulation of high-voltage power equipment on its resistance to lightning voltages, were used. Such critical for the EDT parameters as pulse repetition frequency, energy per pulse, operating lifetime, and resistance of the discharge circuit were insignificant for the VPG intended for high-voltage test laboratories. Work on VPG adaptation to operating conditions of the technological EDT complexes carried out in the High-Voltage Research Institute (HVRI) and in some other institutions gave positive results, but the situation did not change significantly. High-voltage pulse capacitors and spark switches remained weak links. The vicious circle was formed—nobody was ready to invest a lot of money in the development and batch production of new pulsed voltage sources intended for the EDT in the absence of consumers of this technology, and without such investments, it was difficult to bring the EDT to the level competitive with the existing technologies to fulfill needs of potential consumers.
3. Predominantly two liquids habitual and well familiar to high-voltage engineers—liquid hydrocarbons (transformer and diesel oils) and water attractive in many aspects—were mainly used as flushing-out liquids and working environment for well drilling. Hydrocarbon liquids were unsuitable for well drilling because of their high cost and ecological danger. The application of water as a working EDT environment encountered other restrictions, namely, owing to high electrical conductivity and dielectric permittivity, high-amplitude pulses with very steep front should be used to provide discharge channel penetration into the rock. This increases the requirements imposed on the pulse source and insulation of the drilling rig.
4. High operating voltage of EDT installations (200–600 kV for well drilling, approximately 500 kV for reinforced concrete product destruction, 180–400 kV for granulation and disintegration of minerals and concrete, and 200–300 kV for cutting) is a negative psychological factor (fear of high voltage) for service personnel.
5. High price of completing parts.

6. High levels of acoustic noise, electromagnetic radiation, and electromagnetic interference complicate technological installations and require additional means to limit negative impacts.

Among the problems that prevent incorporation of cutting and breaking into industry, in addition to those listed above, are:

- Insulation of electrodes working in strong electric fields under shock loads,
- Very high operating voltage,
- Cyclicity of cutting and breaking that required rearrangement of the equipment to a new position in first stages of the development.

Rapid development of the mechanical methods of rocks destruction and application of new materials reduced the EDT competitiveness and limited introduction of new destruction principles.

Among the *subjective* reasons are:

1. Excelent results of first EDT application gave project managers the illusion of universality of this technology; therefore, they did not pay attention to search and choice of fields of EDT application where it could be competitive with already existing technologies that have already been mastered for many decades. This led to dispersion of forces and means for EDT introduction into the fields where it was obviously noncompetitive: drilling of explosive wells on open-cast mines with fracturing rocks, ring drilling of prospecting wells, and granulation and disintegration of materials having low mechanical strength (for example, coal). Moreover, this discredited the EDT in the opinion of potential consumers.
2. Project managers did not give proper attention to problem of searching liquids that could better fulfill the requirements imposed on flushing-out liquids.

High requirements of the existing and potential consumers of this progressive technology stimulated expansion and deepening of scientific investigations and design elaborations. Their results are presented in the monograph together with achievements of previous decades.

References

1. Emelin MA, Morozov VN, Novikov NP et al (1990) New methods of rock destruction. Moscow: Publishing house Nedra, 240 p (in Russian)
2. Sulakshin SS (2004) Destruction of rocks during drilling of wells. Publishing house of Tomsk Polytechnic University, 136 p (in Russian)
3. Vozdvizhensky BI, Rebrick BM (1989) Into the earth: exploratory drilling from the past to the future. Publishing house Nedra, Moscow, 168 p (in Russian)
4. Aadnoy BS, Cooper L, Miska SZ et al (2009) Advanced drilling and well technology. Society of Petroleum Engineers, USA, 876 p
5. Protasov YI (2001) Destruction of rocks. Publishing house MGGU, Moscow, 453 p (in Russian)
6. Siomkin BV, Usov AF, Kuretz VI (1995) Bases of electropulse destruction of materials. St. Petersburg, Nauka, 276 p (in Russian)

7. Kuretz VI, Usov AV, Zuckerman VA (2002) Electropulse disintegration of materials. Publishing House of the Kola Scientific Center of the Russian Academy of Sciences, Apatity, 324 p (in Russian)
8. Ushakov VY, Dulzon AA (2001) Performance capability of technological installations using the electrical discharge or conductor electrical explosion energy. In: International conference on pulsed power applications. Gelsenkirchen, 27–29 Mar 2001, Pap. Numb. A.03
9. Usov AF, Siomkin BV, Zinoviev NT (2000) Transient processes in installations of electropulse technology. St. Petersburg, Nauka, 160 p (in Russian)
10. Vorob'ev AA, Vorob'ev GA, Chepikov AT Certificate of the opening of NA-122. Regularities of breakdown of a solid dielectric at the interface with a liquid dielectric under the action of a voltage pulse, from 29.04.1998 with the priority of 14.12.1961 (in Russian)
11. Dulzon AA, Ushakov VY (2001) State and development perspective of technological applications of high voltage high-powered pulse. In: International conference on pulsed power applications. Gelsenkirchen, 27–29 Mar 2001, Pap. Numb. A.02

Chapter 2
Impulse Electrical Strength of Rocks and Concretes

As follows from materials presented in Sect. 1.2, the EDT efficiency and practicability are determined by the electrical strengths of destructible/processed solid, surrounding/working media, and insulation/constructive parts of the high-voltage working tool. A large volume of the material on electrical strengths of liquid and solid dielectrics has been accumulated in connection with requirements of high-voltage technology intended for electric and power engineering and military applications [1–4]. The electrical strength of rocks and concretes was studied with a focus on the EDT development; therefore, the most part of the material was obtained in Tomsk Polytechnic University.

Data on the dependence of the electrical strengths of rocks, concretes, and liquid and solid dielectrics on the parameters determining the reliability of the equipment and the EDT efficiency, including physicochemical and mechanical properties and structure of non-conductive materials, voltage impulse duration and shape, thickness of the destructible layer, and the field configuration in the discharge gap, pressure and temperature, and influence of previous breakdowns on the residual electrical strength are presented in Chaps. 2, 3, and 6.

2.1 Physico-Mechanical and Electro-Physical Characteristics of Rocks and Concretes and Their Influence on the Electrical Strength

Objects of EDT destruction (drilling, cutting, breaking, granulating, and disintegrating) can be more than 100 types of rocks. Sometimes drilling is executed under conditions of permafrost with a wide variety of its structures and properties. A wide variety of the nomenclature (by the structures, properties, designs, etc.) is also characteristic for concrete and reinforced concrete products.

V. Y. Ushakov et al., *Electro-discharge Technology for Drilling Wells and Concrete Destruction*, https://doi.org/10.1007/978-3-030-04591-3_2

2.1.1 Rocks

Depending on their origin, rocks can be subdivided into three main groups: igneous, sedimentary, and metamorphic.

The ***igneous rocks*** formed near the Earth surface (effusive rocks)—basalts, porphyries, gabbros, diorites, granites, syenites, etc.—possess the enhanced strength and abrasivity.

The majority of ***sedimentary rocks***—sandstone, quartzite, clay, limestone, marble, etc.—are characterized by lower strengths than igneous ones.

Metamorphic rocks—quartzite, limestone, gneiss, marble, silicon slate, etc.—are formed of rocks of the first two groups; their strengths can be higher or lower than the strength of rocks of the first two groups.

Rocks, as a rule, have complex composition and structure; they consist of a mineral skeleton or a solid phase and a pore space filled with liquid and gas components. Pores are small spaces filled with the mineral skeleton, either closed or interconnected with each other and with the atmosphere. For the EDT, of the greatest interest are such characteristics as porosity, strength, hardness, electrical resistance, and dielectric constant.

The porosity is a property of rock to contain volumes not filled with the solid phase (closed or interconnected with each other—opened). The porosity is usually described by the porosity coefficient in percentage. The coefficients of total and open porosity of various rocks and minerals lie in a wide range. Depending on the mineral structure, conditions of bedding, degrees of diagenesis, age, and structure of deposits or rocks they change from values close to zero to 90%.

In oil-gas-water saturated rocks, the sum of volumes occupied by oil, gas, and water is equal to the total volume of the pore space. The coefficient of closed porosity seldom exceeds 1–5%. Depending on the porosity coefficient K, all rocks can be subdivided into five groups:

(1) highly porous, with $K = 20\%$,
(2) enhanced porous, with $K = 15$–20%,
(3) medium porous, with $K = 10$–15%,
(4) reduced porous, with $K = 5$–10%,
(5) low-porous, with $K = 5\%$.

The effective (average) pore diameter differs significantly for different rocks. Large super capillary pores have the diameter greater than 10 μm. They are typical for pebbles, gravel, coarse and medium-grained sands. Capillary pores with an effective diameter of 10^{-4}–10^{-7} m are typical for more dense sedimentary rocks: fine-grained sands, sandstones, etc. Thin sub capillary pores with a diameter of 10^{-7}–10^{-9} m are characteristic for clays, limestones, and other natural sorbents. The mobility of the liquid phase in the sub capillary pores is minimal. The micropore diameter is commensurable with sizes of molecules; therefore, they are filled with very firmly bound water.

Mechanical strength of the rock characterizes the ability to resist mechanical actions of different intensities and characters without destruction. Among the characteristics of the strength properties of rocks are uniaxial compressive strength, uniaxial tensile strength, and bending strength. The ultimate tensile strength for the majority of rocks does not exceed 20 MPa and is (0.1–0.02) of the compressive strength. The rock strength depends on many factors: mineralogical composition, structure, texture, porosity, hardness of particles, degree of weathering, etc. The sedimentary rocks naturally increase their density with increasing depth of their depth in the Earth crust under the effect of increasing geostatic pressure, mainly at the expense of their reduced porosity. The density increases especially fast at depths up to 500 m. The density of magmatic and metamorphic rocks changes with the depth only slightly.

The mechanical strength of rocks depends on the conditions of application of deforming forces. The least resistance to compression is observed under uniaxial compression. Under biaxial compression, the resistance of rock increases by a factor of 1.5–2.0; it increases even greater under all-round (volume) compression. From this it follows that the conditions of rock destruction in a well in drilling differ from those of strength determined by the method of unilateral or bilateral compression. Wells are penetrated under conditions close to volume compression. The presence of flushing-out liquid column affecting penetration as well as of compaction and hardening of rocks at great depths causes the drilling rate to decrease. One of the important features of rocks is the strength reduction after loading in comparison with undeformed rock, since loading causes irreversible shear deformations of rocks. Because of this, the rock strength decreases.

Fine-grained quartzites and nephrites have the greatest compressive strengths in the range 500–600 MPa. Considerable strengths (greater than 350 MPa) have dense fine-grained granites; gabbros, diabases, and coarse-grained granites have somewhat lower strengths. The greatest tensile strength is characteristic for quartzites and low-porous fine-grained marbles.

By the electrical strength the rocks differ from each other much less, than by the mechanical one. For example, whereas by the compressive strength quartz differs seven folds from sandstone, their electrical strengths differ by less than 2 times, Fig. 1.5. This circumstance is a prerequisite for the applicability of the ED method of destruction of rocks of any mechanical strength and especially high efficiency of its application for drilling or granulating of rocks of enhanced strength.

The hardness of rocks is their resistance to destruction under application of loads of any type. For the destruction of rocks by the mechanical methods, the classification of rocks by hardness is of great importance. The hardness is estimated by the hardness coefficient according to M. M. Protod'akonov scale (f_P) in which all rocks are subdivided into 10 classes with the hardness coefficient from 0.1 to 20 (20 is for quartzites and basalts, 10 is for granites, very hard sandstones, limestones, quartz ore veins, and 1 is for dense clay and coal) [5].

The rock hardness is a complex parameter formed by a number of physicomechanical properties influencing the destruction process in drilling. The hardness is a constant which is independent of the way of drilling.

An important characteristic of rocks is the ***electrical conductivity.*** The rocks have the most contrast electrical conductivity, and this parameter is defined as basic one in the majority of electrical prospecting methods used at present. An important role is played by the electrical conductivity of rocks in ED destruction, in particular, by drilling.

The specific electrical resistance of minerals changes in wide limits from 10 to 10^{10} Ω m. The values of the electrical conductivity and mineral density are interrelated. Minerals with enhanced density also have high electrical conductivity. Since rocks, with few exceptions, represent a set of three phases—solid, liquid, and gas, the mechanism of their electrical conductivity, including electronic, ionic, and mixed conductivity types (more often ionic) is complex. As a rule, the liquid phase (usually mineralized water) whose specific resistance is by several orders of magnitude less than the resistance of rockforming minerals is contained in the rock. Therefore, water saturation is the main factor that determines the electrical resistance of rocks.

For the EDT, of great importance is the ***dielectric constant*** ε of the rock and liquid phase, since the depth of the discharge channel penetration into the rock depends on the relationship between ε values of the solid and liquid. The dielectric constant of minerals is mainly caused by elastic (ionic and electronic) polarization. The relative ε value of the most part of minerals is within the limits 4–12; greater ε values have denser minerals.

The dielectric constant of the large part of rocks is greater than of the minerals, which demonstrates a higher degree of saturation by moisture of rocks in comparison with minerals. The dielectric constant increases with density and is within 5–20. The temperature increase to 200–300 °C causes ε to decrease with the specific density of water. The ε value depends on the mineral composition of the rock, its temperature, and in a less degree on the structural features and pressure.

A study of permafrost with application to the EDT is of great importance, because eternally frozen soils occupy more than 60% of the territory of Russia, and seasonally frozen soils occupy more than 80% of the territory. Considerable areas of such soils are situated in a number of other countries of the northern hemisphere. The thickness of frozen rocks depends on a number of factors and fluctuates from several meters to several hundred meters. In the region of the polar circle, permafrost was detected in the prospecting well to a depth of 1500 m. The average temperature of the eternally frozen rocks in different regions fluctuates from +3.5 to −15 °C. In the region of frozen rocks, a large number of different minerals are concentrated; their exploration and mining requires large volumes of drilling.

Frozen soil represents a complex multicomponent system whose properties depend on interrelations among individual components. The compression and shear mechanical strengths of frozen soil increase with decreasing temperature. The increase of the strength of frozen soils demonstrates hardening of the ice crystal lattice when water impregnating the rock freezes.

Drilling of wells in eternally frozen rocks requires preservation of the natural temperature regime: application of flushing-out liquids with negative freezing temperature preventing warming of the well walls, freezing of drilling tool, freezing of the well, and so on.

Special research is required for ED drilling of eternally frozen rocks. Ice in frozen soils is a cementing material, and the electrical strength of ice is rather high. Investigations of frozen alkaline granite, sandstone, and sand samples and of ice-rock combinations demonstrated that the breakdown voltages (for voltage exposure time of 0.2–10 s) of the strongest (frozen sand) and weakest (frozen sandstone) samples differ two times [6].

In the first case, ice has a decisive effect on the breakdown voltage, and in the second case, it is the rockforming mineral. The electrical strength of frozen sand is close to that of ice. The electrical strength of frozen alkaline granite is by 25% higher than of conventional granite. Hence, the impulse electrical strength of frozen rocks and soils is greater compared to their impulse strength at a positive temperature.

Investigations of electric breakdowns and destructions of more than 100 types of rocks performed mainly at the Tomsk Polytechnic Institute have shown that the breakdown voltage decreases from massive poorly crystallized rocks to weakly cemented sedimentary rocks. The type of liquid surrounding the rock (transformer oil, diesel oil, oil-based solution, water, etc.) has practically no effect on the impulse breakdown voltage of rocks under conditions characteristic for the EDT.

It was established that the hardness coefficient f_P, the tensile strength σ_{br}, density p_0, longitudinal wave propagation velocity C_0, acoustic strength $\lambda_0 = p_0 C_0$, and U_{br} are correlated. For rocks, an increase in the hardness coefficient f_P, elasticity modulus, and tensile strength is accompanied, with rare exception, by the corresponding increase in U_{br}, Fig. 2.1.

In this case, the coefficient of correlation between the breakdown voltage and f_P and also between σ_p and σ_{br} is 0.94 and 0.8, respectively.

In the literature, a number of approximations for U_{br} as functions of the impulse parameters, discharge gap geometry, and properties of rock are suggested:

$$U_{br} = U_0(1 + T/t)^{0.2} \tag{2.1}$$

$$U_{br} = U_{ign} + S_p S/U_{av}, \tag{2.2}$$

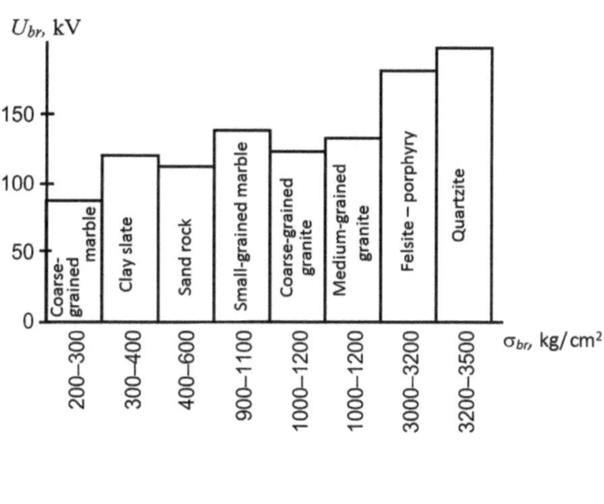

Fig. 2.1 Diagram of distribution of impulse breakdown voltage of rocks versus their compressive strength. Time before breakdown $t_{br} = 10^{-6}$ s; tip-plane electrodes

$$U_{br} = m\, U_0 S_p^{0.2S} S^n \tag{2.3}$$

$$U_{br} = mt + T, \tag{2.4}$$

$$U_{br} = U_0 k S^{0.5}. \tag{2.5}$$

here U_{ign} and V_{av} are the discharge ignition voltage and the average velocity of discharge channel propagation depending on the steepness of the voltage pulse front; U_0 is the minimal breakdown voltage for samples with a thickness of 1 cm, in kV; S is the sample thickness, in cm; T_{br} is the time before the breakdown, in μs; A_p is the steepness of the impulse front, in kV/μs; n, m, k, and T are the coefficients depending on the properties of the destructible material and on the impulse shape.

2.1.2 Concretes

Concrete is called the artificial stone material obtained from the concrete mixture after its formation and solidification. The concrete mixture comprises binders, water, large and small fillers, and in certain cases special additives.

Mineral binders are called powder-like substances that after watering are converted into the plastic dough capable of solidification and petrifaction with time as a result of physical and chemical processes.

As a *fine filler* for concrete, natural and artificial sands obtained by granulation of some rocks, slags, and annealing materials are used. Sand for concrete should comprise grains of different sizes (from 0.14 to 5.0 mm), so that its intergrain voids were minimal; the less volume of voids, the stronger is concrete.

As a *coarse filler* for concrete, gravel or rubble is used. Rubble is subdivided into natural (from a scree) and artificial rubble obtained by granulating of rocks, bricks (brick rubble), and blast-furnace and open-hearth slag (slag rubble). According to grain sizes, the rubble and gravel are subdivided into the following fractions dosed separately when preparing concrete mixtures: 5–10, 10–20, 20–40, and 40–70 mm.

The structure has a decisive effect on the strength and deformation characteristics of concrete. It is roughly nonuniform and depends on numerous factors: grain structure of coarse and fine fillers, volume concentration of cement stone, water/cement ratio, compaction methods, solidification conditions, degree of cement stone hydration, etc. The concrete structure is formed by the spatial cement stone lattice filled with coarse and fine filler grains with numerous micropores and capillaries containing chemically unbound water, water vapor, and air. Therefore, concrete represents a capillary-porous stone material in which integrity is disrupted and all three phases—solid, liquid, and gaseous—are presented.

The structure of the cement stone in concrete is also complex and nonuniform. The cement stone is formed by crystal compounds: crystalline hydrates having complex structure and possessing a very large dispersion and hence, large specific surface (up to 200 m^2/g) and high surface activity. The cement stone comprises elastic crystal structure and viscous mass—a gel—filling it. The combination of elastic and viscous structural components of the cement stone gives concrete elasto-plastic and creeping properties. These properties are manifested in the behavior of concrete under loading and during interaction with the surrounding. The total pore volume in the cement stone under normal solidification conditions is 25–40% of the volume of the cement stone. Capillaries with radii up to 1 μm form 60–80% of the pore volume. The strength and deformation characteristics of concrete are determined experimentally depending on its structure.

The strength theories (of maximal normal strain, maximal tangential strain, etc.) developed for other materials are inapplicable to concrete. Since the strength and deformation properties of fillers and cement stone are different, cracking, especially in the zone of contacts of the cement stone with fine and coarse fillers, can be one of the sources of destruction of this material. In this case, one of the main reasons is low resistance of concrete to the action of tensile strains.

The electrotechnical properties of *concrete* of interest for the EDT depend on its structural components—the filler and the cement stone. In their turn, the electrotechnical properties of the filler (sand, gravel, or rubble) formed by such rocks as granite, marble, diorite, quartzite, limestone, sandstone, and so on depend on the electrophysical properties of rockforming minerals (quartz, feldspar, mica, or calcite). Thus, the electrical strength of concrete with the quartz filler is higher than of concrete with the marble filler, since the contact zone is formed on the quartz surface, which is electrically stronger than on the marble surface. Changes of particle sizes of the coarse filler have no effect on the electrical strength of concrete, but increase its spread. By the present time, a sufficiently large volume of data on the electrical strength of rocks has been accumulated, including the data used for concrete preparation. It was established that the impulse breakdown voltage U_{br} of crystallohydrates is much higher than U_{br} of the cement stone [7]. This demonstrates considerable influence of the discharge processes in pores and capillaries filled with air on the cement stone. A dependence of U_{br} on various parameters of structural porosity of the cement stone, including the pore sizes and total porosity, was established. It appeared that U_{br} of the cement stone having a finer pore structure is higher even in the case in which its total degree of porosity is higher. As a result, the conclusion was drawn that the electrical strength of the cement stone is primarily determined by the size of the through pores and capillaries.

Proceeding to the electrical properties of concrete as a whole, we must note that dry concrete has high dielectric properties. Its bulk electrical resistance changes within the limits (10^4–10^{10}) Ω m; the dielectric loss tangent does not exceed 0.10–0.15; and the dielectric constant is close to 10.

In [8, 9] it was shown that U_{br} for concrete in a sharply inhomogeneous field depends on two factors: electrophysical properties of the filler and properties of the cement stone. Thus, if the electrical strength of the filler is higher than the strength of the cement stone, a breakdown occurs along the cement stone or along the cement stone—filler interface. If the electrical strength of the cement stone is higher, a discharge occurs along the filler. Since the filler, as a rule, has the electrical strength significantly higher than the cement stone, the electrical strength of concrete is mainly determined by the properties of the cement stone.

The electrical strength of concrete is several times smaller than the strength of the filler or of the cement stone, which is caused by the presence in concrete of a contact zone between the filler and the cement stone forming the third structural element (the weak link) along which a breakdown occurs.

A correlation between the mechanical and electrical strengths (though weak) is also observed for concrete. In the system of electrodes located on one free surface, the electrical strength grows with increasing brand of concrete. Moreover, the growth of the breakdown voltage is slowed down with decreasing time of voltage application as the concrete brand B_c increases.

The dependence $U_{br} = f(B_c)$ is observed for decimeter thicknesses of the *rod-plane* system of electrodes located opposite each other. Figure 2.2 shows the volt-time characteristics of concrete of brands 200 and 300 at $S = 10$ cm that confirm the presence of a correlation between the mechanical and electrical strengths of concrete.

For further investigations, as a main parameter describing the physicomechanical properties of concrete, we have chosen its compressive strength that is a general, average characteristic of the material. In addition, the compressive strength serves as the main strength characteristic of concrete.

Considering the effect of many factors on the conditions on which the impulse electrical strength of concrete depends, the mathematical method of multi-factor planning of the experiment was used to derive an expression for calculating the impulse breakdown voltage. Based on the results of the experiments described above, we considered the following factors influencing the breakdown voltage of concrete:

- $\log t_{br}$—logarithm of time before breakdown, from −7 to −6.5; it corresponds to the time period $(0.1–2.5) \times 10^{-6}$ s;
- S—interelectrode distance, 0.03–0.37 m;
- B_c—concrete brand, 100–600.

To convert natural values of the factors into dimensionless values to construct the standard matrix of the experimental plan, the factors were normalized according to Table 2.1. As the basic level, the middle of the interval was chosen for which the

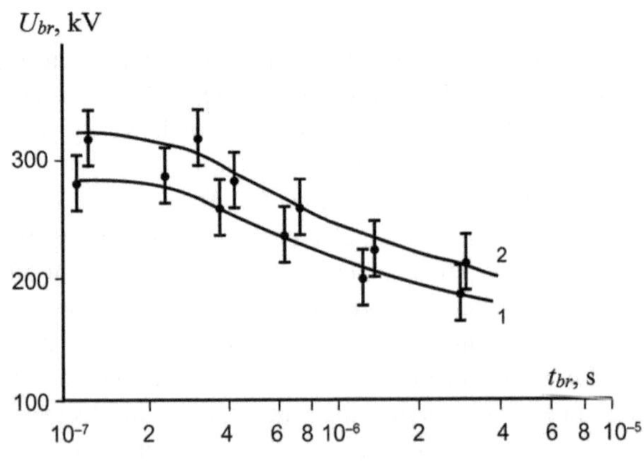

Fig. 2.2 Volt-time characteristic of concrete of two brands (1—200 and 2—300) at $S = 10$ cm

Table 2.1 Normalization of the factors

Range of variation and limiting factor values	Factors		
	Log t	S, m	B_c
Main level (0)	-6.3 (0.5×10^{-6})[a]	0.20	350
Range of variation	0.4	0.10	150
Lower limit (-1)	-6.7 (0.19×10^{-6})	0.10	200
Upper limit (+1)	-5.9 (1.3×10^{-6})	0.30	500
Star shoulder (-1.682)	-7.0 (0.1×10^{-6})	0.03	100
(+1.682)	-5.6 (2.5×10^{-6})	0.37	600
Code mark	x_1	x_2	x_3

[a]Time before the breakdown is indicated in the parentheses

experiment was carried out. To realize the plan of experiment, 20 series of experiences were carried out, and in each series breakdowns of 15–20 samples were initiated. The voltage exposure time required according to the plan was preset in preliminary experiments by adjusting the forming elements in the discharge circuit of the generator. After processing of the accumulated data with exception of insignificant coefficients ($\alpha = 0.05$), from a regression analysis establishing the correlation of the concrete breakdown voltage with the time before breakdown, interelectrode distance, and concrete brand, the following model was constructed:

$$U = -1270.7 - 465.8 \log t - 4649.5 S - 0.2 B_c - 855 S \cdot \log t \\ -37.5 \log^2 t + 670 S^2 + 0.0007 B_c^2 \text{ [kV]}.$$

In this case, the error was 14%.

The model of regression analysis allows the impulse breakdown voltage of building concrete to be determined depending on the concrete brand, RCP thickness, and time before breakdown.

2.2 Influence of the Factors Manifested Themselves in EDT on the Electrical Strength of Rocks and Concretes

2.2.1 Voltage Impulse Duration, Polarity, and Shape

2.2.1.1 Rocks

For *rocks*, as well as for other dielectric materials and media, the law is valid—the electrical strength increases with decreasing voltage exposure time. Only in some special cases and in comparatively narrow interval of impulse durations this law can be violated (for example, see Fig. 2.3). (This law is universal—to achieve identical effect of mechanical action on a system, its magnitude should increase and its duration should decrease.)

For the EDT, another effect is of principal importance, namely, the rate of increase of the electrical strength of rocks and concretes (as, indeed, of other non-conducting solids) is less than of liquids. When the voltage exposure time decreases by two orders of magnitude (from 10^{-5} to 10^{-7} s), the breakdown voltage of the majority of rocks increases only twice, which is much less than the breakdown voltage of liquids (see Sect. 3.2.1).

By the present time, the dependences of the breakdown voltage of rocks on the impulse duration for the breakdown at the front of an oblique-angled voltage impulse

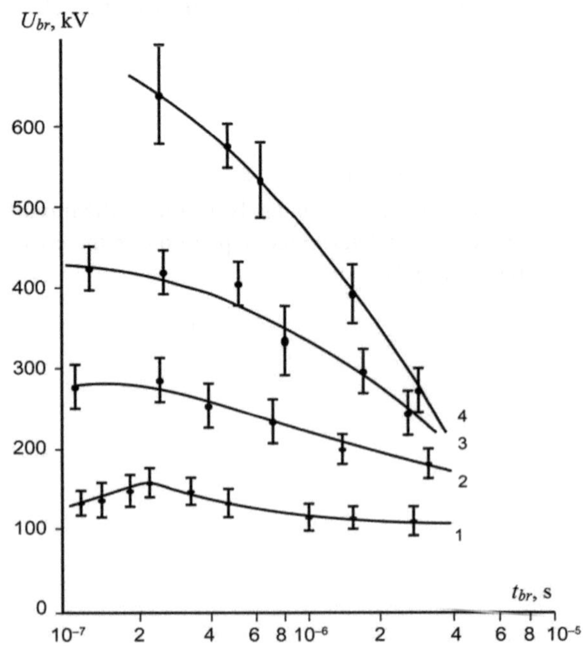

Fig. 2.3 Voltage-time characteristics of concrete of brand 200. Interelectrode distance $S = 5.3$ (1), 10 (2), 20 (3), and 50 cm (4)

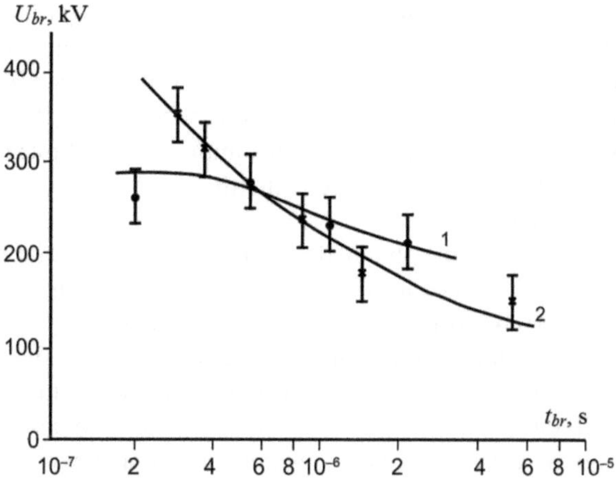

Fig. 2.4 Voltage-time characteristics of concrete of brand 250 (1) and of technical water (2); $S =$ 10 cm

or a plane rectangular impulse with nanosecond front have been determined and described by empirical equations.

The majority of studies on the breakdown of rocks were performed for the tip-plane electrode system characterized by the least values of the breakdown voltage with other conditions remaining the same. This provided the least breakdown voltage and simplified experiments. Since electrodes in drilling and cutting are mounted on one surface of a solid, the data on the breakdown voltage obtained for the tip-plane electrode system can be taken only as tentative when choosing the EDT regime.

2.2.1.2 Concretes

Regular investigations of the voltage-time characteristics of concrete were performed for samples with thickness of 5–80 cm. Mass produced concrete of brand 200 was investigated. Breakdown was initiated at the front of a single pulse of positive polarity in the +R −P system of electrodes located on the opposite sides of the sample and placed in the discharge chamber filled with transformer oil. In this case, the depth of surface impregnation of samples with transformer oil did not exceed 2–3 mm. The breakdown time (t_{br}) changed from 10^{-7} to 3×10^{-6} s. Figure 2.4 shows the VTC of concrete of brand 200 with the thickness $l = 5.3$, 10, 20, and 50 cm.

The behavior common for all dielectric media—an increase in the electrical strength with decreasing voltage exposure time and its decrease with increasing thickness of the destructible layer—can be seen.

Thus, for example, for $S = 10$ cm, $t_{br} = 3 \times 10^{-6}$ s, the breakdown gradients E were (16–20) kV/cm, while at $t_{br} = 10^{-7}$ s, they were (25–30) kV/cm. (The gradients for concrete are much less than the corresponding values for rocks.)

As well as for drilling, to choose the impulse parameters of technical devices for RCP destruction, it is necessary to compare the VTC of concrete and technical water.

Figure 2.4 shows the dependences $U_{br} = f(t)$ for concrete and technical water illustrating the scatter of breakdown voltages. In experiments, samples of concrete of brand 250 with sizes of 10 × 10 × 10 cm were studied manufactured by the Tomsk Plant for Large-Panel Housing Construction from the concrete mixture used for RCP production. The breakdown of samples and technical water with the resistivity $\rho = 6 \times 10^5\ \Omega$ m was initiated using the rod-plane electrode system with interelectrode distance of 10 cm.

Analyzing the obtained dependences, it should be noted that the tendency toward a decrease in the rate of increase of the breakdown voltage at $t = 10^{-7}$ s (see Figs. 2.2 and 2.4) confirmed in experiments with concrete of brand 250 (Fig. 2.4) is a favorable factor for the choice of the impulse parameters of the voltage source. As can be seen from Fig. 2.4, the interval of voltage impulse durations causing the breakdown in the balk rather than on the surface of the solid is somewhat wider for concretes in comparison with rocks. Thus, the concrete breakdown probability of 100% corresponds to the voltage exposure time shorter than 0.3×10^{-6} s. For voltage exposure time in the range $(0.3–1) \times 10^{-6}$ s, the breakdown of both concrete and water is possible.

The voltage impulses with the above-indicated durations provide destruction of concrete of brands up to 400 using the electrode system shown in Fig. 2.5a. For the electrode system shown in Fig. 2.5, the conditions of concrete breakdown were much worse. However, this arrangement of electrodes arose in the final stage of RCP destruction, when the product surface was extensively destructed and covered by a network of cracks. Therefore, there is a probability of concrete breakdown sufficient for complete destruction of the RCP. To destruct RCP made of concrete of brands 500–600, the voltage exposure time should be decreased to 0.2×10^{-6} s. It should be emphasized that in practice of recycling of substandard RCP such cases are expected to be seldom, since the majority of products subject to recycling are made of concrete of brands 100–400.

When the VPG is connected to a low-resistance load (in this case—to the electrode system in technical water), the impulse amplitude decreases, and the source voltage should be increased to the value $U_0 = U_{br}/K_{red}$, where $K_{red} = 0.5–0.8$ is the voltage reduction factor. The impulse amplitude required for the destruction of concrete of

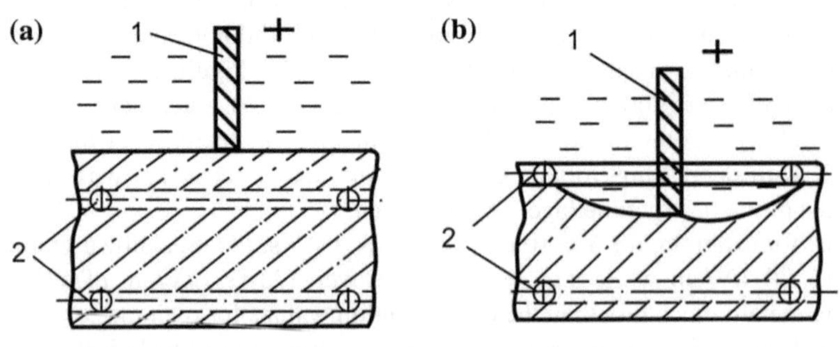

Fig. 2.5 Variants of arrangement of the high-voltage electrode relative to the armature. Here 1 denotes the high-voltage electrode and 2 denotes armature

various brands and thicknesses can be calculated from Eq. (2.2) taking into account the decrease on the low-resistance load.

2.2.2 *The Interelectrode Distance and the Position of the Electrodes Relative to the Material or Article to Be Destroyed*

2.2.2.1 Rocks

In breakdown of rocks and other dielectric materials and media, ***the effect of total voltage***—nonlinear (saturating) relationship of the breakdown voltage with the ***interelectrode distance***—is manifested. Moreover, the effect is more pronounced for porous and mechanically less strong rocks.

The dependence of the breakdown voltage on the interelectrode distance is described by the empirical equation of the form

$$U_{br} = kS^{m} \text{ [kV]}, \tag{2.6}$$

where S is the distance between the electrodes, m is the exponent ($m = 0.5$–0.7 for $S = 10$–60 mm and $m < 0.5$ for $S > 100$ mm), k is the coefficient considering the rock type, the breakdown condition, and the statistical property of the breakdown voltage (deviation of each concrete measurement from the average value [10–12]).

Numerous data on the breakdown of solid dielectrics and rocks demonstrate that the breakdown voltages have statistical properties that can be described by a normal distribution law.

It was established that the breakdown voltage of rocks was influenced significantly ***by the arrangement and configuration of the electrode system***: in case of rock breakdown and rod-rod electrode system located on one surface of the sample, the breakdown voltage increased by 10–25% (depending on the rock type) in comparison with the rod-rod electrode system located on the opposite surfaces of the sample.

To increase the EDT efficiency in drilling and cutting of rocks, it is necessary to provide the highest probability of channel penetration into the rock. To this end, a maximal breakdown voltage $U_{\max}$ obtained in the study of electrodes applied and maximal impulse amplitude should be used under repeated action of voltage impulses to initiate a breakdown at the impulse front. The difference between values of the maximum breakdown voltage and the breakdown voltage averaged over the indicated number of measurements for a wide range of sample thicknesses is illustrated by Fig. 2.6 [13].

Investigations were performed using the VPG with adjustable values of the voltage and deposited energy. The electrode system consisted of two rod electrodes with a diameter of 10 mm. The electrodes were placed on one surface of a granite sample. The distance between the electrodes S changed in the range 10–300 mm. Granite had a medium-grained structure with a relatively narrow range of variations of the specific mechanical compressive strength $\sigma_{com} = 120$–160 MPa. The sample sizes were 400 × 300 × 250 mm. The liquid medium was polluted by transformer oil with electrical strength $E = 40$ kV/cm.

The average and maximal values of the breakdown voltage for granite can be described by the power-law dependence (2.7):

$$U = U_{10} S^m, \tag{2.7}$$

where $U_{10} = 200$ and 224 kV are breakdown voltages for interelectrode distance $S =$ 10 mm and average (U_{av}) and maximal ($U_{\max}$) values of the voltage, respectively; $m = 0.3$ is the experimental coefficient; S is given in cm. Curves 1 and 2 in Fig. 2.6 are the calculated maximal (curve 1) and average (curve 2) values of U_{br} for different S [13]. For rock salt and ice, $m = 0.6$, and for microquartzite, $m = 0.4$.

Hence, the dependence $U_{br} = f(S)$ remains in the form $U_{br} \equiv S^m$ for the breakdown of rocks having different mineralogical composition with radically different electro-physical characteristics (for example, rock salt, sandstone, and microquartzite).

2.2.2.2 Concretes and Reinforced Concrete Products

In the breakdown of concrete, the effect of total voltage is manifested approximately in the same way as in other dielectrics. As can be seen from Fig. 2.7, an increase in the thickness S of the samples from 10 to 80 cm leads to a decrease in E_{br} from 16 to 8 kV/cm.

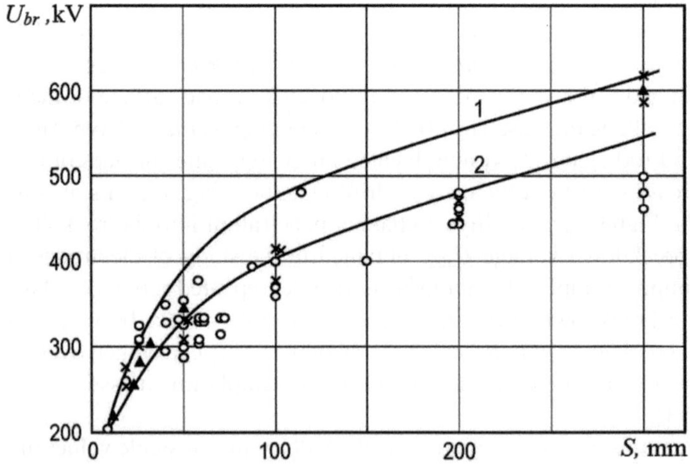

Fig. 2.6 Experimental (symbols) and calculated dependences (solid curves) of the breakdown voltage on the interelectrode distance for the number of impulses $N = 3$ (empty circle), 6 (times), and 6–15 (filled triangle); curve 1 is for U_{max}, and curve 2 is for U_{av}

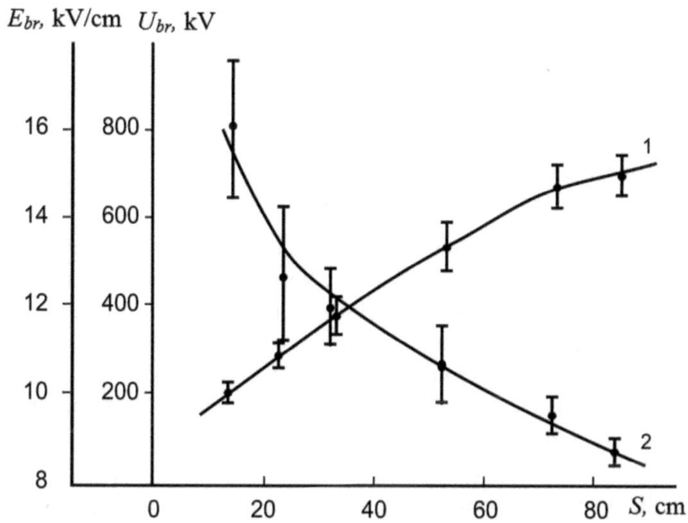

Fig. 2.7 Breakdown voltages (1) and penetration field gradients (2) for concrete of brand 300 depending on the sample thickness (The slope of the pulse front is 1050 kV/μs)

The effect is explained by two reasons: for all dielectric materials and media, the increase of the inhomogeneity of the electric field in the interelectrode space with increasing distance between the electrodes; for liquid and solid dielectrics, one more effect is manifested—an increase in the number of inhomogeneities statistically distributed in the dielectric (weak regions) and hence, an increase of the breakdown probability for them at decreased field gradients.

For ED technologies of RCP destruction, the electrode system is expedient in which the high-voltage electrode is located on the surface of the product being destructed and the armature, all elements of which are electrically connected in already in the stage of product manufacture is served as the earthed electrode. In such electrode system when the RCP surface is not broken and the armature near the high-voltage electrode is not bare (Fig. 2.5a), the application of an impulse leads to channel formation in the thickness of concrete (naturally, in the event that the impulse amplitude is sufficient for the breakdown). At the initial moment, the character of the electric field in such electrode system is close to that in the rod-plane (R-P) system. In the process of RCP destruction, the interelectrode distance decreases from several tens of centimeters to zero, the electric field approaches to that of the system rod opposite rod or to the field of the rod-rod system located on one surface. Thus, for the ED technologies of RCP destruction, the upper estimation of the impulse amplitude should correspond to the rod-plane system for which the interelectrode distance is maximal.

When the armature near the high-voltage electrode is bared (Fig. 2.5b), the probability of RCP surface flashover is high. Such mode in the ED technology is not working, the efficiency of concrete destruction in this case is significantly lower than of the breakdown in the solid thickness.

2.2.3 Temperature and Static Pressure

High EDT potentialities revealed in the earliest stages of the development of this technology gave supervisor, professor A. A. Vorob'ev grounds to formulate in 1960 the problem of EDT adaptation for solving the problem of drilling of deep and super deep wells. (Even by the present time, no principal obstacles for solving this problem by the EDT have been found.) The first results of works in this direction were presented in [9, 14]. The electrical strength of rock samples in transformer oil was considered in these studies (for a time before breakdown ~0.6×10^{-6} s) separately for changes of the pressure and temperature. This only partially reproduces conditions in a deep well, since physicomechanical and electrophysical properties of rocks in the bowels are determined by the joint influence of high pressures and temperatures.

For the tip-plane electrode system (in the case of the through breakdown), the impulse breakdown voltage of rock samples in transformer oil with a hole under the tip electrode increases by 20–30% when the external pressure increased from atmospheric one to 15 MPa. The pressure increase above 5 MPa does not lead to the increase of the breakdown voltage. This fact can be explained by liquid penetration into porous rock samples, since for lightly porous dielectric (fluoroplastic and vinyl plastic) materials, the breakdown voltage is independent of the pressure. In the same works, results of investigations of rock (quartzite, felsite-porphyry, marble, clay slate, or sandstone) breakdown depending on the temperature in the range 20–140 °C were presented. The breakdown voltage remained invariable up to a temperature of 80 °C, and then decreased for all materials. Based on these data, the conclusion was drawn that the increase in temperature up to 140 °C has practically no effect on the impulse breakdown voltage of rocks and will not worsen the destruction condition at these temperatures.

The assumption was put forward that the increase of the breakdown voltage of porous rock samples with pressure is determined mainly by their impregnation with surrounding liquid, which in drilling plays the role of the working (flushing-out) liquid. In this case, the highest relative increase in the breakdown voltage is observed for the rock with a higher porosity (in this case, for marble). In drilling of wells, both water and oil-gas saturated rocks are encountered whose degree of saturation affects the breakdown and destruction processes.

It was natural to assume that the experimental results obtained under laboratory conditions would depend on the procedure of preliminary impregnation of rock samples with a liquid. The impregnation of rocks (marble or sandstone) with different liquids (water, glycerine, transformer oil, or ethyl alcohol) within 24 h increased their breakdown voltage by two and more times, and this increase was the greater, the higher was the porosity of rock and the dielectric constant of the liquid. Here the analogy is pertinent with the breakdown of the combined (paper-oil or paper-film-oil) insulation widely used in high-voltage instrument making (capacitors, cables, inputs/bushing insulators, etc.). Careful impregnation increases the electrical strength of the combined isolation up to a factor of 2.5.

In the 90s and zeros, scientific investigations of the joint influence of the elevated voltage and temperature on the electrical strength and destruction were continued at the High-Voltage Research Institute (HVRI) to determine the possibility of effective drilling of wells to depths of 3000–4000 m. These investigations were supported by Schlumberger Ltd. (Grant No. REO-1360-TO-02).

All experiments were performed in VersaPro drilling solution with sandstone samples (σ_{com} ~ 60 MPa), limestone (σ_{com} ~ 120 MPa), and granite (σ_{com} ~ 160 MPa). The same samples for each rock were cut from one stone. In connection with laboriousness of experiments and manufacture of samples, their number in each examined series was no more than 10. The choice of rocks was caused by radically different physicomechanical characteristics and composition and their wide spread in nature.

In addition, sandstone and granite as a material for samples and the VersaPro drilling solution as a medium into which the samples were put were recommended by the customer. Investigations were performed in a specially developed and manufactured metal chamber with a volume of 6300 cm^3, Fig. 2.8.

An impulse with amplitudes up to 400 kV was supplied to the chamber through the high-voltage input made of polycarbonate (lexan) whose working temperature could reach 150 °C. Samples were broken down at the front or amplitude of the

Fig. 2.8 Test stand comprising test chamber 1, cap 2, input of high-voltage insulator 3, high-voltage screen 4, high-voltage cable 5, and filter 6

voltage pulse. The rod-rod electrode system with rod diameter of 10 mm and flat face was used.

The distance between the electrodes was 20 mm. The electrodes were placed on one surface of the sample. In all experiments the samples were held in the drilling solution for 2 h. The rate of sandstone impregnation by the VersaPro drilling solution was preliminary determined. At a temperature of 17 °C and pressure $P = 5$ MPa, it was 0.19 mm/min, at $P = 20$ MPa it was 0.2 mm/min, and at $P = 25$ MPa it was 0.25 mm/min. For rock breakdown the penetration depth did not exceed 0.32 s [8]; in our case, it was 6.4 mm. Hence, the depth of the discharge channel was always less than the impregnation depth. This event was periodically tested. To this end, samples were sawn along the discharge channel trajectory.

The electrical strength of the tested rocks quickly increased as the pressure increased up to 5 MPa, Fig. 2.9. With further increase in the pressure, the electrical strength remained practically unchanged. Analogous results were obtained by other authors under different experimental conditions, including the tip-plane electrodes (the through breakdown) and the transformer oil medium.

It should be noted that the dependence $E_{br} = f(P)$ for the impregnated rocks, Fig. 2.9, was analogous to that for the VersaPro drilling solutions.

This confirms the conclusion on the determining influence of impregnation of porous rocks on their electrical strength. The porosity of the examined rocks differed significantly. The average porosity of granite is approximately 3.5%, limestone—5.6%, and sandstone—12%. Figure 2.10 illustrates the dependence of the relative change of the electrical strength of solid materials E_{20}/E_{01} on their porosity, where E_{20} is the electrical strength at $P = 20$ MPa and E_{01} is the electrical strength at $P = 0.1$ MPa. Polyethylene was taken as a non-porous material for which E_{br} was independent of the pressure up to $P = 35$ MPa. Its water mass absorption is 0.01%, that is, water in it can be present only in the form of molecules, and such viscous liquid, as the drilling solution, does not penetrate into its volume.

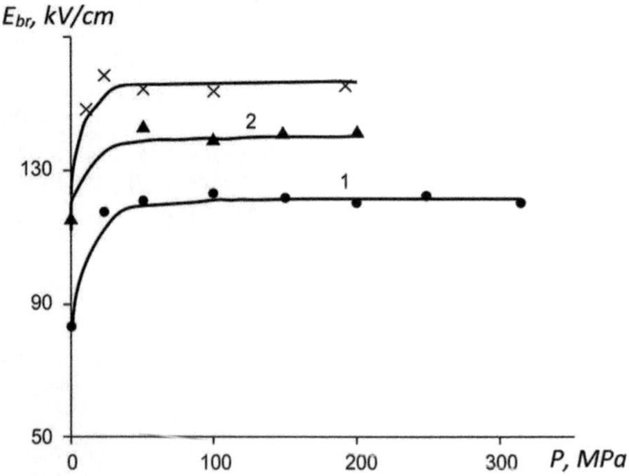

Fig. 2.9 Dependence of the electrical strength of rocks on the pressure for sandstone (1), limestone (2), and granite (3)

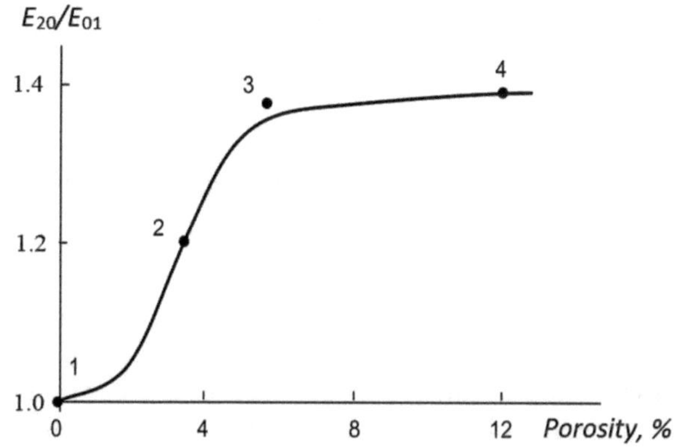

Fig. 2.10 Dependence of the relative electrical strength on the rock porosity for polyethylene (1), granite (2), limestone (3), and sandstone (4)

From Fig. 2.10 it can be seen that the relative impulse electric strength increases with porosity especially fast for the porosity less than 6–7%. The dependence in Fig. 2.10 confirms the effect of strengthening of porous rocks caused by the impregnating liquid. For impulse durations characteristic for the EDT, the electrical strength of the liquid (VersaPro drilling solution) was higher than of rocks.

Figure 2.11 shows the dependences of the electrical strength of rocks and VersaPro drilling solution on the joint influence of the pressure and temperature. The character of the dependences of the electrical strength for rocks with the electrodes placed on one sample surface approximately corresponds to the analogous dependence for the drilling solution. At a pressure of 30 MPa and a temperature of 100 °C, the electrical strength of the examined rocks approaches the electrical strength of the drilling solution: the difference makes 3.2–12.8%, whereas in the initial state, that is, at a pressure of 0.1 MPa and temperature of 17 °C, it reaches 16.5–27.7%.

The joint influence of the elevated pressure and temperature has a great effect on the electrical strength of rocks with less porosity (increasing it), especially at relatively low pressures ($P < 10$ MPa) and temperatures ($T < 50$ °C), Fig. 2.12. From Fig. 2.12 it can be seen that the reverse dependence of the relative electrical strength E_{max}/E_{01} on the porosity in comparison with Fig. 2.10 is observed. This can be explained by the fact that the intensity of sample impregnation in the presence of small pores increases under the joint influence of the pressure and temperature. The structure of the porous space depends on the rock type and metamorphism.

Thus, under the joint influence of the elevated pressures and temperatures on the examined rocks, the character of changing their impulse electrical strength is ambiguous. At the pressure $P > 10$ MPa and temperature $T > 50$ °C, the electrical strength increases to a maximum faster for samples with smaller porosity and smaller pore sizes. Further increase in P and T caused the electrical strength to decrease. It approached the electrical strength of the impregnating liquid (drilling solution) being the electrically strongest component in the combination rock—impregnating liquid.

Here the liquid component has the determining influence on the impulse electric strength of the combination. As the temperature exceeds a certain value depending on the pressure, its further increase inevitably leads to the decrease of the electrical strength of the liquid.

2.2.4 Impact of Dynamic Loads Resulting from Preceding Discharges

In ED drilling the rock monolith is influenced by a large number of voltage impulses and breakdowns; therefore, cracks are formed in it. They doubly influence the breakdown voltage, namely, they decrease the average penetration voltage, but in this case, the length of channels of the subsequent breakdowns increases, thereby increasing the breakdown voltage. The two factors counteract; at present they can be considered

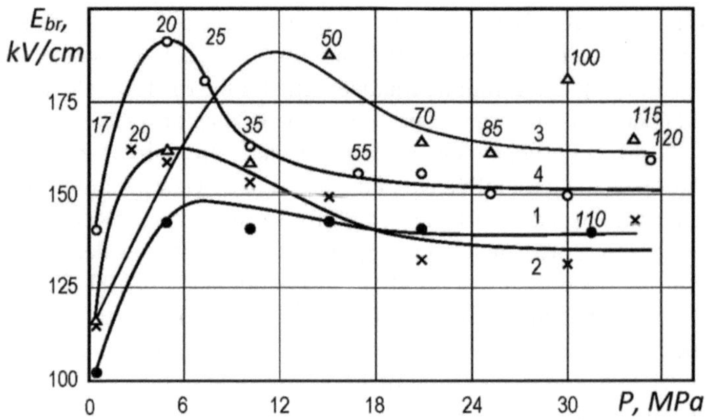

Fig. 2.11 Dependence of the electrical strength of rocks on the pressure and temperature for sandstone (1), limestone (2), granite (3), and drilling solution (4) (Figures near the points indicate the temperature)

Fig. 2.12 Dependence of the relative electrical strength of rocks on the porosity

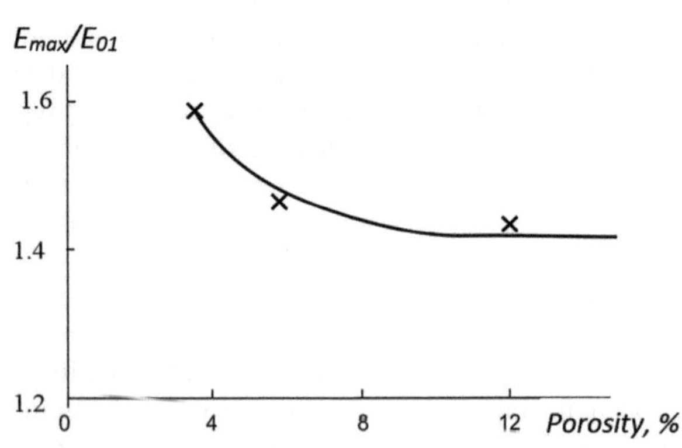

only experimentally. Of practical interest is a comparison of the data on multipulse and single-pulse strength of rocks, Table 2.2.

For comparative analysis, the following quantities were chosen:

All results of Table 2.2 were reduced to the distance between the electrodes $S = 120$ mm,
As a reference breakdown voltage, $U_{br} = 486.3$ kV was chosen obtained in well drilling in granite with drilling head having a diameter of 360 mm for the electrode spacing $S = 120$ mm,
As the beginning breakdown voltage U_{beg}, $U_{br} = 630$ kV was accepted, obtained in breakdown of granite by the first pulse.

From Table 2.2 it follows that investigations with granite gave statistically close values of the decrease in the breakdown voltage (15.9–25%) and deviation of the breakdown voltage from drilling one (2.9–9.0%).

The effect of previous impulses on U_{br} is maximal when initial and stabbing craters are formed in the rock massif.

Table 2.3 presents experimental results, where U_{beg} and U_{edge} are the breakdown voltage at the beginning of destruction and at the edge of the stabbing crater (i.e., in the region of the strongest mechanical failure), respectively; $(U_{beg} - U_{edge}/U_{beg}) \times 100$ is the relative decrease in the breakdown voltage, in %; t_{br} is the time before the breakdown, in μs; W_s is the energy deposited per unit length of the discharge channel, in J/mm.

The results obtained demonstrate that for close times before breakdown, the breakdown voltage decreases at the edge of the slabbing crater U_{edge}, reaching 18% at $S = 40$ mm. Slowing down of the relative decrease in U_{edge} with increasing S is caused by an increase in the size of the slabbing crater and a decrease in the energy deposited into the breakdown channel that affect the length of growing cracks when energy is liberated in the discharge channel [15]. This suggests that the decrease in U_{br} at the edge of the initial slabbing crater is caused primarily by the formation of macro- and micro-cracks in the undestructed volume of the rock after beginning of destruction.

In a multi-electrode system, for example, in a drilling head, the discharge in each individual event of electric impact on rock is initiated from the electrodes over the

Table 2.2 Comparative characteristics of granite and microquartzite breakdown

Method	Drilling, $S =$ 120 mm	Edge of the slabbing crater	Allowance for cyclic motion of the electrodes	Multiple breakdown and destruction
Working U_{br}, kV	486.3	533.8	515–530	472.1
Deviation of U_{br} from the initial value, %	22.8	18.0	15.9–18.2	25.0
Deviation of U_{br} from the drilling voltage, %	0	9.7	5.9–9.0	2.9

Table 2.3 Characteristics of granite breakdown at the beginning (U_{beg}) and at the edge of the slabbing crater (U_{edge})

S, mm	40	50	60	70
U_{beg}, kV	328.6	319.7	309.5	331.0
U_{edge}, kV	269.3	285	287	315
$(U_{beg} - U_{edge}) \times 100$	18.0	10.8	7.3	4.8
t_{br}, μs	0.18	0.21	0.21	0.22
W_s, J/mm	14.4	11.5	10.2	9.3

entire bottom, but the breakdown, as a rule, is completed in one electrically least strong site. In this case, numerous channels of incomplete breakdowns are formed at the bottom under the electrodes, thereby weakening the electrical strength of rock due to the cumulative effect of sequential breakdowns of rock. The surface layer integrity is broken: it is covered by a system of micro- and macro-cracks that also decrease the breakdown voltage. The breakdown voltage may be halved due to the impact of previous discharges from many hundred impulses.

Numerous attempts to consider this effect were undertaken when developing technological regimes of drilling. For example, in [16] U_{br} was investigated for repeated granite breakdown when the electrode spacing changed from 100 to 140 mm. The experimental conditions were the following:

Pulse front duration of 0.35 μs;
Energy stored in the VPG changed in a range 6–7 kJ depending on the working pressure;
Energy deposited into the discharge channel changed within the limits 41.3–35.0 J/mm;
The breakdown voltage and the time before breakdown were measured for each impulse;
Breakdown was always excited at the amplitude or front of the voltage pulse with minimal overvoltage;
Granite samples with sizes (500–600) × (400–500) × (350–400) mm having compressive mechanical strength of ~140 MPa were put in transformer oil;
Two or three fixed rod electrodes with diameter of 10 mm were placed on one surface of the sample;
For each electrode system, 6–7 experiments were performed; the results obtained were averaged during subsequent processing.
The EDT suggests the application of multi-electrode destructive tool (drilling head, actuator in the process of RCP destruction, etc.).

In drilling wells, the number of discharge gaps in the penetration region can be different depending on the design of the destructive tool. For example, 12 discharge gaps were used for the drilling head with a diameter of 360 mm and electrodes spaced at $S = 120$ mm. The process of electropulse drilling is cyclic. The cycle duration is determined by the number of impulses n_c necessary for destruction of the entire penetration region. As a result, the drilling implement reached a cyclic depth of h_c, and the cycle was repeated.

The average impulse number (n_c) necessary for one cycle of a multi-electrode system can be described by the empirical equation:

$$n_c = \frac{0.785 \cdot D_{well}^2 \cdot h_c}{v_o \cdot \psi} \text{ [impulses]}, \tag{2.8}$$

where D_{well} is the well diameter, in cm; h_c is the depth reached in one cycle, in cm; v_0 is the volume productivity of the impulse, cm^3/impulse; and ψ is the probability of discharge penetration in the rock. The depth h_c depends only on the electrode spacing:

$$h_c = 1.53 + 0.1S. \tag{2.9}$$

The cycle duration increases with decreasing electrode spacing. Depending on the design features of the multi-electrode destructive tool and the set of energy characteristics of the impact impulse, the number of impulses in the cycle can largely fluctuate, reaching several hundred impulses. For individual pair of electrodes, the number of impulses in one cycle, on average, can reach ten and even more.

With increasing number of impulses, the destruction crater depth and width increase. (The influence of different factors on the parameters of slabbing crater is studied in Chap. 5). This causes the discharge channel length to increase. Figure 2.13 shows the dependence of the maximal granite destruction depth on the number of impulses for $S = 100$ mm for a two- and three-electrode systems.

The destruction depth fast increased when the number of impulses increased to a certain value (~20 impulses), then deepening practically stops, that is, the discharge

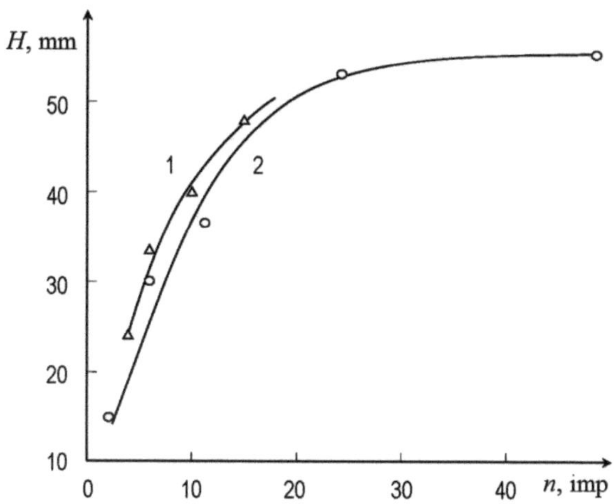

Fig. 2.13 Dependence of the destruction depth on the number of impulses for two-electrode system (1) and three-electrode system (2)

Fig. 2.14 Dependence of the breakdown voltage on the number of impulses for the two-electrode system (1) and three-electrode system (2)

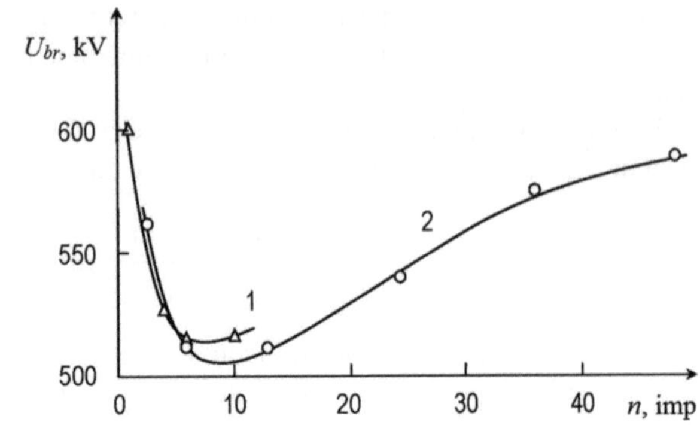

channel does not further penetrate into granite. Since the length of the discharge channel at the maximal depth reached 160 mm, that is, was by a factor of 1.6 greater than under initial breakdown conditions, the flashover was observed, as a rule, at the surface of the formed crater or in oil between the electrodes along the shortest path.

Figure 2.14 shows the dependence of U_{br} on the number of acting impulses at $S = 100$ mm for the two- and three-electrode systems.

For the first ten impulses, the breakdown voltage decreases by 15.8%.

Further increase in the number of acting impulses leads to an increase in the breakdown voltage up to its initial value. From the viewpoint of technological application, it is very attractive to choose the working voltage equal to the minimal breakdown voltage corresponding to the application of 7–10 impulses. In this case, the destruction depth reaches 30–40 mm, Fig. 2.15.

Calculations by Eqs. (2.8) and (2.9) demonstrated that for the employed drilling tool, the average duration of the cycle (the number of impulses) for one pair of electrodes $n_{c1} \cong 10$ impulses, and the penetration depth for the cycle is $h_c = 13.5$ mm.

Fig. 2.15 Dependence of the breakdown voltage on the destruction depth for $S = 100$ (curves 1 and 4), 120 (curve 2), and 140 mm (curve 3). Here curves 1–3 are for the two-electrode system, and curve 4 is for the three-electrode system

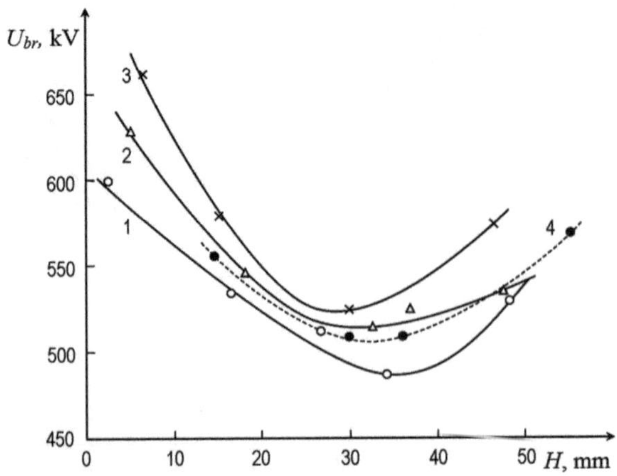

From Fig. 2.13 it can be seen that upon exposure to 10 impulses, the maximal destruction depth is 35–40 mm, and for $S = 120$–140 mm, the destruction depth will be greater, that is, deepening of the drilling tool per cycle is smaller than the maximal destruction depth. This is caused by considerable roughness of the well bottom: protrusions and tips are observed at the bottom. They do not allow the drilling head to deepen at the maximal destruction depth. Hence, under conditions of the above-described experiments, after the end of the cycle and deepening of the drilling head, tips with depths of 21.5–26.5 mm are formed, which can determine the choice of the working voltage. Figure 2.15 shows the dependences of U_{br} on the maximal destruction depth for two- and three-electrode systems at $S = 100$, 120, and 140 mm. The minimal breakdown voltage is observed for all S in the range of destruction depths $H = 30$–35 mm. Based on Fig. 2.15, taking into account the depths of the remaining tips, the working pressure at $S = 120$ mm should be no less than 530–515 kV rather than 630 kV, as for the initial conditions, that is, by 15.8–18.2% less.

In actual drilling by the above-indicated drilling head, with time before breakdown 0.30–0.32 μs, the maximal value of U_{br} was 486 kV (see Chap. 7). Taking into account the inhomogeneity of the internal structure of granite samples, the agreement between the penetrative (working) voltages can be considered acceptable (the deviation was 8.3–5.6%).

In this case, it is most likely that U_{br} decreased to a minimum due to the increase of the degree of macro- and micro-destructions in the rock volume. The increase in U_{br} with further increase of the number of acting impulses was caused by the predominant effect of the increase of the discharge channel length.

References

1. Ushakov VY (2004) Insulation of high-voltage equipment. Springer, Berlin, Heidelberg, 421 p
2. Ushakov VY, Klimkin VF, Korobeynikov SM (2007) Impulse breakdown of liquids. In: Ushakov VY (ed), Springer, Berlin, Heidelberg, 397 p
3. Vorob'ev AA, Vorob'ev GA (1966) Electrical breakdown and destruction of solid dielectrics. Publishing house Vysshaya Shkola, Moscow, 224 p (in Russian)
4. Vershinin YN (2000) Electron-thermal and detonation processes in the electric breakdown of solid dielectrics. Publishing house of Uro RAN, Ekaterinburg, 259 p (in Russian)
5. Emelin MA, Morozov VN, Novikov NP et al (1990) New methods of rock destruction. Publishing house Nedra, Moscow, 240 p (in Russian)
6. Dazkevich SY, Gafarov RR, Lopatin VV et al (2006) Pulsed electrical strength of liquids and solid at high pressure. Izvestia vuzov. Physika; Appl (11):484–486
7. Krivitsky EB, Apostoli VL, Sorochinsky AP (1963) Conductivity of the spark discharge channel in water. Electronnaja obrabotka materialov (6):34–37 (in Russian)
8. Kutchinsky GS, Lysakovsky GG, Monastirsky AE et al (1987) The electrical strength of liquid dielectrics under impact of a microsecond pulses. Electrichestvo (10):41–44 (in Russian)
9. Rompe R, Weizel W (1944) Uber das Toeplersche Funken-gesetz. Z Phys 122(3):636–639. (In German)

10. Ushakov VY, Dulzon AA (2001) Performance capability of technological installations using the electrical discharge or conductor electrical explosion energy. In: International conference on pulsed power applications. Gelsenkirchen, 27–29 Mar 2001, Pap. Numb. A.03
11. Burkin VV, Makarov PV, Siomkin BV and others (1975) To calculation of the pressure field around a spark in solid dielectrics. Zhurnal technitcheskoy phyziki 45(11):2395–2399 (in Russian)
12. Blaznin BS, Shchegolev IA, Lozin LI et al (1983) Processing of natural stone by electric discharges. Electronnaja obrabotka materialov (1):5–7 (in Russian)
13. Kuretz VI, Usov AF, Zuckerman VA (2002) Electropulse disintegration of materials. Publishing House of the Kola Scientific Center of the Russian Academy of Sciences, Apatity, 324 p (in Russian)
14. Zel'dovich YB, Riser YP (1963) Physics of shock waves and high-temperature hydrodynamic phenomena. Publishing house Gosudarstvennoe Izdatelstvo physika – matematitcheskoy literature, Moscow, 632 p (in Russian)
15. Siomkin BV, Usov AF, Kuretz VI (1995) Bases of electropulse destruction of materials. St. Petersburg, Nauka, 276 p (in Russian)
16. Occhini E (1971) A statistical approach to the discussion of the dielectric strength in electric cables. IEEE Trans Power Apparatus Syst 90(6):2671–2682

Chapter 3
Liquids Used as an Insulation and a Working Medium

3.1 Requirements for Liquids Used in the EDT

The EDT imposes a number of rigid requirements on liquids that perform three functions: of a working medium in contact with the body being treated (destructed), insulation in the transmission line (the drilling implement), and flushing-out liquid (removal of drilling cuttings). As a working medium and insulation, liquids should possess:

- high electric strength,
- low electric conductivity,
- high thermochemical resistance to the action of electric discharges and fast recovery of properties after the breakdown.

The experience of EDT application for insulation of various liquids that has been accumulated by the present time demonstrates that depending on the circumstances, the dielectric liquids (transformer and diesel oils, etc.), water, and special flush-out liquids meet the requirements listed above.

- Compositions of various solutions with petroleum and oil base and water were developed at the HVRI with participation of the Moscow Institute of Petrochemistry named I. M. Gubkin (MIP).

Their application decreases only slightly the ED drilling performance characteristics in comparison with drilling in diesel fuel.

Nowadays new non-polluting electrically insulating liquids and drilling solutions based on petroleum are used in the world for drilling of oil and gas wells, including:

- Midel 7131 transformer liquid (see Technical Datasheets Nos. 3 and 7 of 2007),
- Versa Pro and Versa Clean drilling solutions based on oil (Schlumberger Ltd) used for drilling of deep wells for oil and gas,
- Calc-bitumen solution (CBS) developed at the MIP.

V. Y. Ushakov et al., *Electro-discharge Technology for Drilling Wells and Concrete Destruction*, https://doi.org/10.1007/978-3-030-04591-3_3

Table 3.1 Main electrophysical characteristics of liquids

Liquid	Midel	Versa Pro	Versa Clean	CBS	Transformer oil	Tap water
ρ, Ω cm	5×10^{12}	2.9×10^{8}	4.7×10^{8}	6.0×10^{9}	10^{12}	$(4–5) \times 10^{3}$
ε	3.2	8.3	7.2	3.8	2.2	81
E_{br}, kV/cm	135	115.0	139.5	95	110	78
t_{br}, μs	0.3	0.35	0.36	0.3	0.3	0.4
Standard punch U_0, kV	62	11	12	34	40	0
ρ_0, g/cm^3	0.97	1.12	1.16	1.15	0.88	1.0

Table 3.1 presents the main electrophysical characteristics of the above-indicated liquids, transformer oil most widely used at early stages of EDT development, and water used to destruct the reinforced concrete products (RCP) and to granulate the filler and suitable for drilling wells under certain conditions.

In Table 3.1, the following designations are used: ρ is the volume resistivity; ε is the relative dielectric constant; ρ_0 is the relative density; E_{br} is the average impulse breakdown strength for initial liquids $E_{br} = U/S$; and t_{br} is time before the breakdown. All liquids have sufficiently high electric strength and approximately identical times before the breakdown, but differ significantly by the resistivity. The Versa Clean solution turns out to be the best. Its composition is given in Table 3.2.

Table 3.2 Composition of drilling solutions

Versa clean		CBS	
Composition	Volume	Composition	Volume
BDC-99 mineral oil, L/m^3	662	Diesel fuel	50.0
VG-Plus, kg/m^3	8.55	Highly oxidized bitumen	13.2
10-min dispersion		Quicklime	30.0
Lime, kg/m^3	17.11	Sulphonol	1.0
Versamul, L/m^3	19.2	Water	5.8
Versacoat, L/m^3	6.4		
20-min dispersion			
Brine, L/m^3	279.8		
Versatrol, kg/m^3	11.4		
30-min dispersion			
Versamod, L/m^3	3.1		

The Midel transformer liquid is used as an alternative to mineral oils in transformers and other high-voltage equipment. The peculiarity of Midel 7131 is its sufficiently fast biodegradation. According to the data of the supplier, the Midel 7131 liquid biodegraded by more than 89% for 28 test days, that is, it is practically environmentally friendly. This advantage has allowed the HVRI to obtain the permission of the Norway Nature Protection Services to carry out electrodischarge drilling of granite bedrock using Midel 7131 near Bergen in 2009. The data presented in Table 3.1 demonstrate that this liquid is not worse than the transformer oil by a number of characteristics and even surpasses it by environmental and some other indicators.

Versa Pro and Versa Clean drilling solutions. In accordance with the contract with Schlumberger Ltd, 4 drilling solutions: two K23 and K24 water solutions and two Versa Pro and Versa Clean oil-based solutions were tested to elucidate the solution that meet most closely the ED drilling requirements. As a result of tests, the preference was given to Versa Pro and Versa Clean solutions.

The calc-bitumen solution (CBS) was tested as a flushing-out liquid in 1989–1991 in ED drilling of the mine shaft with a diameter of 2 m. The electrophysical CBS characteristics provided by the customer ("Torezantratsit" Industrial Association nowadays belonging to Ukraine) were investigated. Table 3.1 presents its basic characteristics, and Table 3.2 gives its composition. The CBS has the characteristics comparable to those of the polluted transformer oil and can be used for ED drilling. It was impossible to continue these tests because of the collapse of the USSR and discontinuation of these works in Ukraine.

3.2 Influence of Factors Manifested Themselves in the EDT on the Electric Strength of Liquids

A study of the electrophysical properties of liquids allows one to choose the liquids that best meet the needs of the EDT. As already noted above, there are many works devoted to the study of the behavior of various liquids under the action of impulse voltage. They are generalized most fully in [1, 2].

Below we consider only those aspects of the complicated phenomenon—the electric breakdown of liquids—that are directly connected with the EDT development and application.

3.2.1 *Voltage Impulse Duration, Polarity, and Shape*

For liquid dielectrics, in particular, strongly polar ones, the characteristic effect of polarity—higher values of the breakdown voltage for the negative polarity of the electrode determining the non-uniformity of the electric field in the discharge gap—is vividly pronounced. The difference changes from units of percent to several times.

The effect of polarity gave us grounds to suggest that the working high-voltage impulse in the EDT should have the negative polarity. However, the practical importance has not only the value of the working voltage determined by the breakdown voltage of the discharge gap, but also the destruction volume—the volume of a slabbing crater determined by the extension of the discharge channel and the depth of its penetration into the solid. According to the data of numerous investigations performed in the process of the EDT development, the technological efficiency was higher for the positive polarity of the potential electrode; therefore, the overwhelming majority of works on the destruction and drilling of rocks were subsequently performed for the positive polarity of the potential electrode.

The EDT discharge channel is typically formed at the voltage impulse front. The special features of the effect in which the channel formation and the electrohydrodynamic processes in it were completed at the plane part of impulses with nanosecond front were also investigated [3]. Such regime of the electric discharge can also be used in the EDT. In experiments [4], the parameters of the voltage impulse generated by an idling VIG were the following: impulse front duration $\tau_f = 0.15 \times 10^{-6}$ s and half-decay impulse duration $\tau_{imp} = 50 \times 10^{-6}$ s. The electrode system was formed by two rod electrodes with flat ends 10 mm in diameter with a length of 150 mm located at an angle to each other, Fig. 3.1.

No less than 10 breakdowns were performed at each voltage step, and the average breakdown voltage U_{br} and the time before breakdown t_{br} were determined. The water resistivity ρ was periodically controlled in each voltage step.

Figure 3.2a shows the voltage-time characteristics of water with $\rho = 4.1 \times 10^3$ Ω cm for the breakdown at the voltage impulse decay for the electrode system shown in Fig. 3.1 with interelectrode gap lengths $S = 10$–90 mm. An increase in the time before breakdown leads to the fact that VTCs for different S get closer together, that is, the breakdown voltage approaches to the discharge ignition voltage in water. It is well known that the discharge channel ignition voltage in a liquid depends weakly on the interelectrode gap changing in wide limits (1–17 cm) [2]. The dashed curves in Fig. 3.2a show the times of 100% water breakdown for all S. When S changes 9 times, the voltage of 100% breakdown probability changes by 55 kV or by a factor of 1.44. The time of 100% breakdown probability changes from 4.0×10^{-6} to 6.4×10^{-6} s or by a factor of 1.6. Taking into account large variance of times before breakdown, this change can be considered insignificant.

Contrary to the popular belief, a decrease in the electrical conductivity of water by its very careful purification is not an effective method of increasing the electric strength of insulating structures [2].

Figure 3.2b shows the VTCs of breakdown of water for the indicated resistivity and $S = 20$ mm. From curves 1 and 2 it can be seen that the breakdown voltage for the same time before breakdown increases with decreasing resistivity of water. This effect is primarily caused by larger deformation of the voltage impulse (the increased rate of voltage decay) because of a decrease in the electrode system resistance in water (discharge resistance). The literature data on the influence of ρ on the electric strength of liquids with enhanced conductivity are inconsistent. The electric strength (breakdown voltage) of liquids with increased conductivity can be determined not

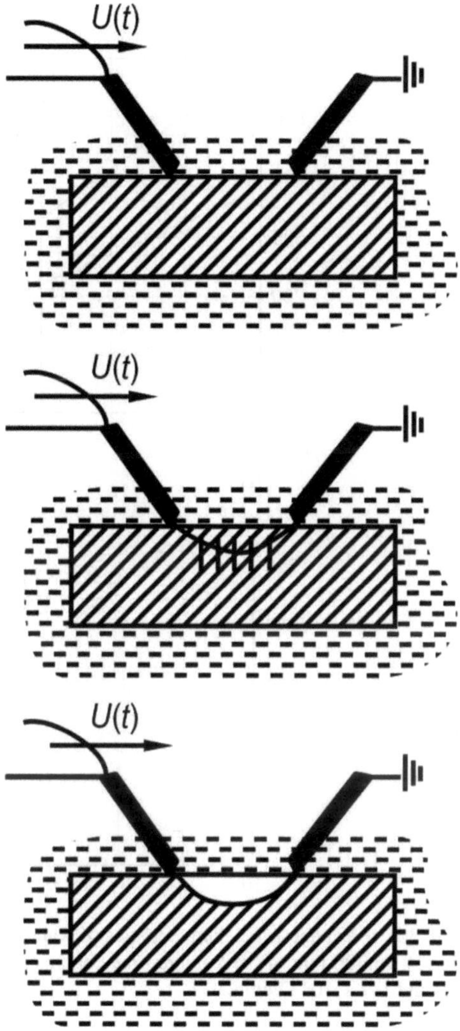

Fig. 3.1 Sequence of breakdown and destruction processes in the system with one free surface

so much by the physical processes that proceed in the discharge gap, as by the parameters of impulse deformation that depend on ρ and on the parameters of the discharge circuit.

It should be noted that the time of 100% breakdown probability of water is practically independent of its resistivity for ρ changing from 1.8×10^3 to 4.1×10^3 Ω cm characteristic for tap water in most regions of Russia.

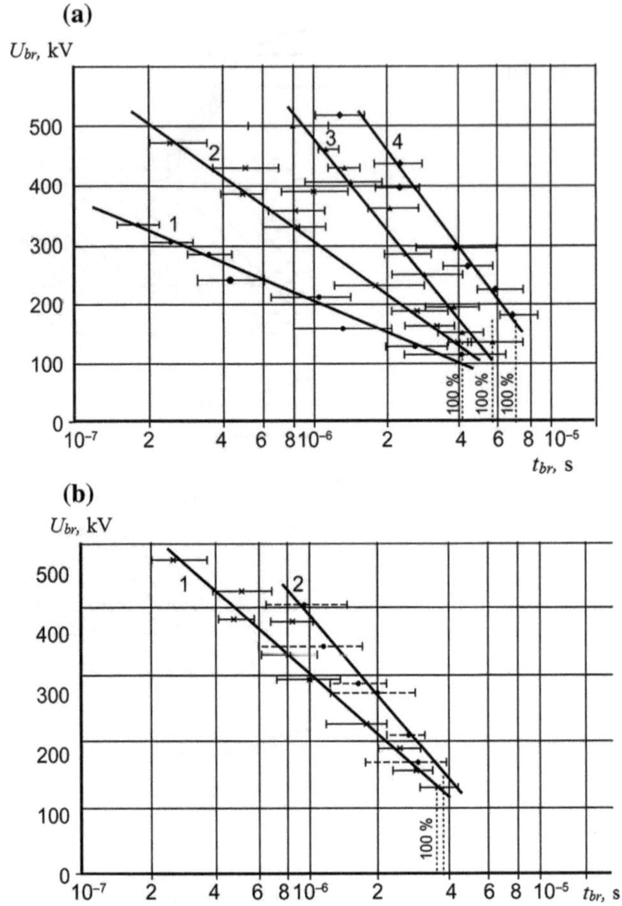

Fig. 3.2 **a** Voltage-time characteristics of water with $\rho = 4.1 \times 10^3$ Ω cm for $S = 10$ (1), 20 (2), 50 (3), and 90 mm (4). **b** Voltage-time characteristics of water for $S = 20$ mm and $\rho = 4.1 \times 10^3$ (1) and 1.8×10^3 Ω cm (2)

3.2.2 Pressure and Temperature

Despite practical incompressibility of the condensed media, under certain conditions the electric strength of liquids and solids increases with pressure [1]. Natures of this effect in breakdown of liquid and solid dielectrics are different. The most probable reason for liquids is a decrease in the influence of gas bubbles being formed or already existed in the liquid in the stages of the discharge channel ignition and evolution [1, 2, 5–7]. These conclusions were drawn under experimental conditions (pure liquids and short discharge gaps with uniform field distribution) that differed significantly from those realized in the EDT. Under these conditions, the double increase in E_{br} was observed when P increased to 15 MPa (an average time before breakdown of 2×10^{-7} s) [1].

Investigations were performed under conditions close to those of the EDT: the liquids contained impurity with high concentration, and the discharge gaps of centimeter length were with sharply non-uniform field [8]. Investigations were carried out in the metal chamber having a volume of 6200 cm^3. The VIG with rated voltage of 500 kV and discharge capacity of 0.01 μF was used as a high-voltage source. The impulse front duration was $\tau_f = 0.25$ μs. The characteristic VIG resistance was 22.3 Ω. These VIG parameters meet the EDT requirements [3]. The temperature in the chamber changed from 17 to 120 °C using heaters located in the chamber. It was measured with a certified thermocouple. The test chamber and the hydraulic system allowed pressure up to 40 MPa to be created and maintained at temperature up to 120 °C. Ranges of pressure and temperature variations corresponded approximately to those observed when the well depth increased up to 3500–4000 m. The breakdown was always initiated at the front or amplitude of the voltage impulse with minimum overvoltage. The impulse voltage parameters in the breakdown were measured with an accuracy no worse than ±3.5%.

To create a sharply non-uniform electric field, a tip-plane electrode system was chosen. The electrodes were made of stainless steel with a plane radius of 50 mm and a tip radius of 0.25 mm tapered at an angle of 30°. Transformer oil, Versa Pro drilling solution, and water were tested. Most experiments for all liquids were performed with positive polarity of the tip electrode. For water, investigations were also performed for negative polarity and in a uniform field.

The average breakdown strength of transformer oil and water in a sharply non-uniform field for positive polarity of the tip electrode did not change when the pressure increased up to 35 MPa for times before breakdown (0.15–0.30) $\times 10^{-6}$ s (curves 4 and 5 in Fig. 3.3). For water, the same was also observed for negative polarity and uniform field (curves 1 and 2).

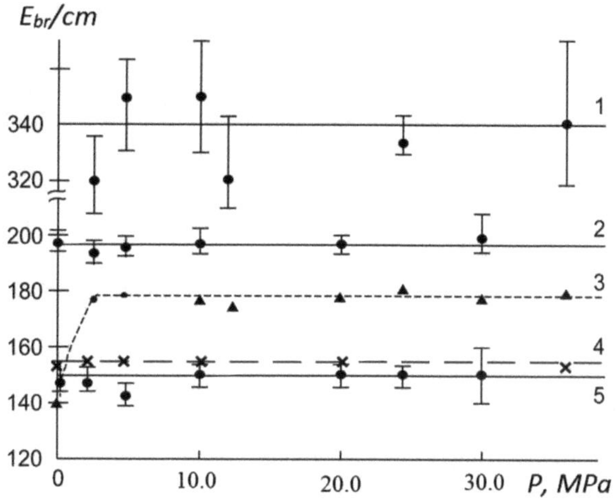

Fig. 3.3 Dependence of the electric strength of liquids on the external pressure. Here 1 is for water, negative polarity, and uniform field; 2 is for water, positive polarity, and uniform field; 3 is for the drilling solution, positive polarity, and tip-plane electrode system; 4 is for oil, positive polarity, and tip-plane electrode system; and 5 is for water, positive polarity, and tip-plane electrode system

As well as for the breakdown at atmospheric pressure, the effect of polarity was observed for the breakdown of transformer oil at a pressure of 15 MPa and times before breakdown less than 0.6×10^{-6} s.

The breakdown strength of Versa Pro drilling solution increased by a factor of 1.26 when the pressure increased to 2.5 MPa and then remained practically unchanged up to pressure of 35 MPa (curve 3 in Fig. 3.3). Such behavior of $E_{br} = f(P)$ can be explained by the peculiarity of the solution composition and structure. It represents emulsion of more than 10 components of solid and liquid substances in the oil medium. Emulsification promotes saturation of the solution by the air-gas phase distributed both in the liquid and solid phases. Long sedimentation does not remove it from the liquid. Special degassing of the drilling solution was not performed. It seems likely that the bubble mechanism of discharge ignition sensitive to external pressure is most probable for this liquid for exposure times of several fractions of a microsecond [2].

Most experiments on elucidation of the influence of pressure and temperature on the electric strength of liquids were performed with the Versa Pro drilling solution, since it is used for mechanical drilling of oil and gas wells, but is not investigated as applied to the EDT.

During the experiment not only the breakdown voltage, but also the electric resistance of the drilling solution was measured for the tip-plane electrode system (with interelectrode gap $S = 20$ mm) depending on the pressure and temperature. Measurements demonstrated that the change of the pressure in the range 0.1–35 MPa at a temperature of 17 °C had no effect on the low-voltage resistivity of the drilling solution. This is in agreement with the results of numerous previous works. The resistivity of any liquid depends significantly on the temperature, since the ion mobility and the degree of thermal dissociation of molecules of the liquid itself and impurities it contains increase with temperature. The resistance R of the drilling solution measured in the experiment under joint influence of the pressure and temperature quickly decreased when the temperature increased up to ~50 °C (curve *2* in Fig. 3.4). Nevertheless, at $T = 120$ °C the resistivity of the drilling solution remained high ($R = 1.85$ M Ω) and sufficient for its use in the EDT.

As applied to drilling of deep wells, of great interest is the joint influence of the temperature and pressure on the impulse electric strength of a drilling solution. For technical liquids containing impurities of different types, the electric strength slowly decreased when the temperature increased up to the value close to the boiling temperature under the action of the impulse voltage.

Curve 1 in Fig. 3.4 shows the change in E_{br} of the drilling solution under the simultaneous influence of the elevated pressure P and temperature T.

It can be seen that the increase of pressure up to 5 MPa and temperature up to 20 °C causes E_{br} to increase by a factor of 1.37. Since the temperature change in the narrow range (from 17 to 20 °C) has practically no effect on the electric strength of liquid dielectrics [9], the noticeable increase in E_{br} should be due to the increase of the pressure. With further increase of the pressure alone, the electric strength of the solution does not change, whereas the simultaneous increase of the pressure and temperature up to $P = 35$ MPa and $T = 120$ °C caused E_{br} to decrease by a factor of 1.29 in comparison with its maximal value. This decrease can be explained by the change of the liquid state at the increased temperature: the decrease of its viscosity, density, solubility of gases, and approach to the phase transition (boiling). Nevertheless, the electric strength of the drilling solution under these conditions meets the EDT requirements.

3.2.3 Interelectrode Gap

It is well known that the manifestation of the scale effect—the decrease of the electric strength with increasing gap length (the effect of the total voltage), the electrode areas, and the volume of the insulation medium within the range of action of the electric field—is common for all dielectric media. The nature of this effect was considered in a number of works (for example, see [1, 2]).

It should be noted that the dependence of the electric strength of liquids on the electrode shape and size, being of great importance for the design of high-voltage equipment, is less important for the EDT, because here such characteristics as the resistance of the insulation and metal parts of the executive elements to the influence of shock loads during discharges and motions, erosive and abrasive wearing, and electric breakdowns of insulation are more important.

When designing technological installations, the interelectrode gap and the position of electrodes relative to the interface between the liquid and the treated (destructed) solid have a more significant impact on the breakdown voltage of liquids, because they influence the design of the actuating element, its reliability, and service lifetime.

The distances between the electrodes, their arrangement relative to each other, and surfaces of the destructed/treated material characteristic for the EDT differ significantly from those for which most of the materials on the electric strength of liquids

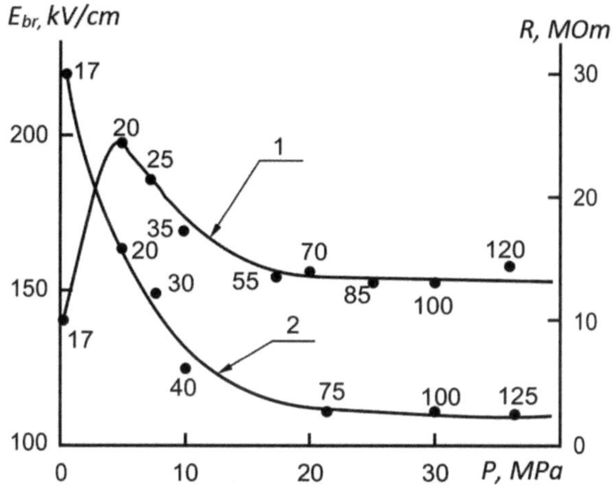

Fig. 3.4 Dependences of the electric strength (1) and resistance (2) of the Versa Pro drilling solution on the joint influence of the pressure (on the scale of abscissas) and temperature (points near the curves)

are presented in the literature. In this regard, to develop installations for EDT and to update them subsequently, careful selection of the data available from the literature and special investigations focused on the EDT needs were required.

In the general form, the dependence of the breakdown voltage on the gap length for all dielectric media can be written as $U_{br} = BS^n$, where B and n are the coefficients depending on the properties of the dielectric medium and the breakdown conditions, and d is the gap length. Moreover, n can be set equal to 1 only for gap lengths changing in relatively narrow limits. The range of gap lengths is $n < 1$ for the EDT. Only for submillimeter gaps the dependence of U_{br} on d is not so unambiguous.

As an example, Fig. 3.5 shows the dependences of the electric strength (E_{br}) of transformer oil and distilled water on S in gaps with a uniform field obtained by different authors for submicron voltage impulse duration. For impulses of microsecond duration, the dependence of E_{br} of distilled water on d in the gaps with a uniform field is shown in Fig. 3.6 [10].

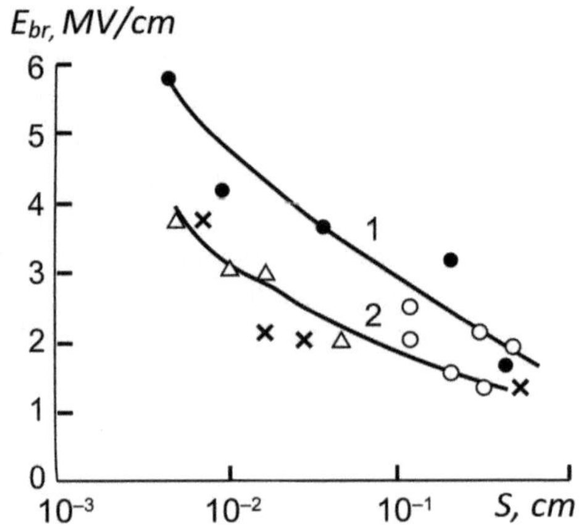

Fig. 3.5 Dependences of E_{br} in a uniform field on d for transformer oil (1) and distilled water (2) with a discharge time of 10 ns

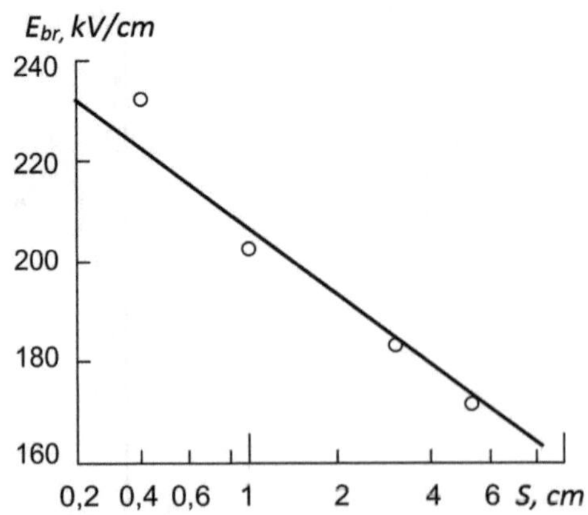

Fig. 3.6 Dependence of E_{br} of distilled water on S for microsecond discharge time in a uniform field

3.2.4 Position of the Electrodes Relative to the Liquid-Solid Interface

The electric strength of liquids at the interface with insulation of the actuating element determines the reliability of its operation and its service lifetime, and at the interface with rock or concrete it determines the feasibility of the discharge channel penetration into the solid and the practicability of the EDT idea itself. When water and other liquids with increased electrical conductivity and dielectric constant (called water below) are used as working media in the EDT, the already complicated problem of coordination of insulation functions of three dielectric media: liquid, solid insulation, and solid being destructed becomes even more complicated. Smaller values of the surface flashover voltage (U_{fl}) in comparison with the breakdown voltage in the liquid and solid volumes, observed in most cases, are due to the manifestation of a number of effects: field distortion due to different dielectric characteristics of the solid and liquid in contact with it, influence of the sorbed substances contained in the liquid and solid; and energy supply of the developing breakdown channel through the solid capacity. The greatest distortion (strengthening) of the field takes place in the triple point—point of contact of the electrode, liquid, and solid. Because of manifestation of the above-indicated effects, U_{fl} on the solid surface is determined by many factors: parameters of the acting voltage (primarily by impulse duration), shape of the solid surface and its position with respect to the electric field lines, character of the electrode contact with the solid, and properties of the solid and liquid in contact with it.

The rate of voltage rise on the impulse front is so important parameter that at large dU/dt (short front duration), the weakest link (concerning the electric breakdown) appears the solid (rock or concrete) rather than the interface or the liquid. In order that this situation was not observed for the insulation of the executive element, special measures should be undertaken (see Chap. 6).

All variety of positions of the solid surface relative to the electric field lines in the study of laws of flashover and measurements of U_{fl} can be reduced to three variants, Fig. 3.7.

In actual high-voltage structures, the electric field pattern is much more complicated because of wide variety of insulator and electrode shapes and their mutual positions. Attempts to transfer the traditional methods of coordination of oil-filled device insulation to the design of EDT insulation using water as an insulator and working medium led to unacceptably short service lifetime and small probabilities of discharge channel penetration into the destructed body.

Attempts to solve this problem using mathematical and physical modeling of the electric field (with application of electrolytic baths) were also unsuccessful. The labor-consuming experimental investigations oriented at the EDT needs and the results of investigations carried out to develop high-voltage electrophysical installations of other designations were required.

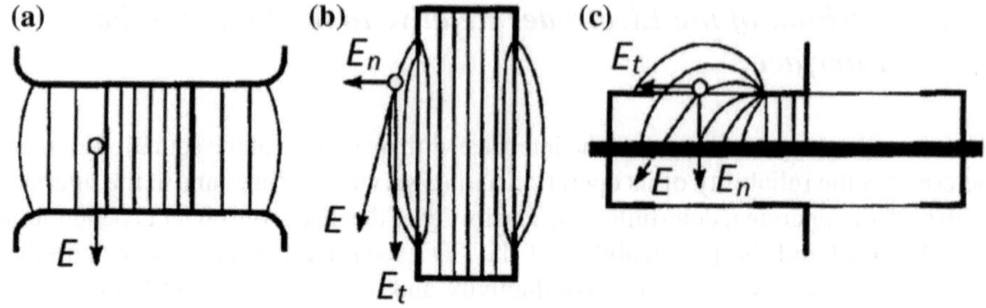

Fig. 3.7 Characteristic variants of positions of the solid surface relative to the electric field lines. Here **a** is for a uniform field and **b** and **c** are for non-uniform fields with prevalence of the tangential (E_t) and normal (E_n) components of the electric field strength vector (relative to the solid-liquid interface)

Based on the results of experimental investigations and the experience accumulated during the development and operation of working bodies of EDT installations with water as a working medium, the following conclusions can be drawn:

1. Field regulation as a method of increasing the discharge voltages of insulating structures is ineffective for insulators operating in water. This method was approved for the following laboratory samples:

a. Insulators of the simplest shape with the surface parallel to the electric field lines,
b. Insulators with the complicated surface providing the field component normal to the surface and the increased flashover length,
c. Insulators in which the field strength in the region of contact with the electrode is decreased by means of:

- external shielding,
- internal shielding,
- rounding insulator edge.

The field regulation engenders the electric field component normal to the insulator surface that decreases the flashover voltage. In most cases, the optimal insulator shape is the simplest one: cylinder, flat washer, etc.

2. The discharge gradients on the surface of insulators in water depend to a lesser degree on the properties of the insulator material than in gases and nonpolar liquids and are determined to a large extent by the electric strength of the water itself.
3. The choice of the material for insulators working in water should be determined by the mechanical strength, the resistance of the surface to the impact of the discharge channel, and the through-breakdown strength.

References

1. Ushakov VY (2004) Insulation of high-voltage equipment. Springer, Berlin, Heidelberg, 421 p
2. Ushakov VY, Klimkin VF, Korobeynikov SM (2007) Impulse breakdown of liquids. In: Ushakov VY (ed), Springer, Berlin, Heidelberg, 397 p
3. Siomkin BV, Usov AF, Kuretz VI (1995) Bases of electropulse destruction of materials. St. Petersburg, Nauka, 276 p (in Russian)
4. Vajov VF, Kozlova NV (2012) Impulse electric breakdown of water and granite. Electrichestvo (6):73–76 (in Russian)
5. Hebner RE, Kelley EF, Fitzpatrick GJ (1987) The effect of pressure on streamer inception and propagation in liquid hydrocarbons. In: 9th International conference on conductor and breakdown in dielectric liquids, № 4, pp 27–39
6. Fuhr J, Aschwanden T (1987) Electric breakdown in highly pressurized water. In: International conference on conductor and breakdown in dielectric liquids, № 4, pp 411–415
7. Gournay P, Lesaint O (1994) Evidence of the gaseous nature of positive filamentary streamers in liquids. In: Conference on electrical insulation and dielectric phenomena. Annal report, № 4, pp 834–839
8. Yutkin LA (1986) Electrohydraulic effect and its application in industry. Publishing house Maschinostroenie, Leningrad, 253 p (in Russian)
9. Dazkevich SY, Gafarov RR, Lopatin VV et al (2006) Pulsed electrical strength of liquids and solid at high pressure. Izvestia vuzov. Physika; Appl (11):484–486
10. Kutchinsky GS, Lysakovsky GG, Monastirsky AE et al (1987) The electrical strength of liquid dielectrics under impact of a microsecond pulses. Electrichestvo (10):41–44 (in Russian)

Chapter 4
Discharge Channel as a "Working Tool" in EDT

After short-circuiting of the electrodes by the discharge channel, the energy accumulated in the VIG starts to be liberated in it. In this stage, the discharge channel acts as a converter of the electric energy of the source into the internal energy of products in the channel, transformed into work on channel expansion, energy of mechanical stress and strain fields, and energy of the newly formed solid surface. The quantitative destruction characteristics are determined by the process of energy transfer to the channel plasma, shock wave generation by the expanded channel, wave dynamics considering the presence of the solid-liquid interface, and formation of stress-strain state of the medium leading to cracking.

The channel stage of the discharge and the subsequent processes of solid destruction can be subdivided into two big groups:

(1) Evolution of the plasma channel: release of the Joule heat in the channel and expansion of the channel;
(2) Mechanical and acoustic processes in the volume of a solid near the channel, including:

Motion of an ensemble of a great number of radial cracks in the circular destruction zone,
Increase in the number of cracks and duration of this stage with simultaneous increase in the energy deposition rate,
Sharp decrease of the number of moving cracks (the stronger, the lower is the energy release rate). However, the rate of growth of mobile cracks in this stage, as well as in the first two, remains maximum,
Preservation of several single mobile cracks. Duration of this stage depends essentially on the energy release rate: for low rates, this stage is practically absent, and for high rates, the cracks grow considerably.

Let us consider the phenomena and the regularities of these two groups of events in a solid in final stages of technological EDT process.

V. Y. Ushakov et al., *Electro-discharge Technology for Drilling Wells and Concrete Destruction*, https://doi.org/10.1007/978-3-030-04591-3_4

It is possible to detect a chain of successive groups of the system parameters each of which is determined by the preceding group, the properties of the destructed/processed material, and the electrode system. As a first approximation, we can consider the following groups of parameters: electric discharge circuit characteristics, energy characteristics of the spark channel, thermodynamic and hydrodynamic parameters of the spark, characteristics of the stress and strain field in the solid, and the parameters characterizing the intensity and character of destruction of the solid. Exactly theoretical and experimental investigations of the above-indicated system parameters and their relationships are the general problem of studying post-breakdown phenomena in solid dielectrics.

4.1 Electrophysical Parameters and Dynamics of the Discharge Channel

By the present time, the majority of the characteristics of the channel stage of solid dielectric breakdown are estimated for a very limited number of dielectric materials and energy deposition regimes. Some parameters of the channel stage obtained by generalization of results obtained by different authors are presented in Table 4.1 from [1].

Table 4.1 Characteristics of the channel stage of solid dielectric breakdown

Breakdown channel characteristic	Dielectric			Reference
	Quartzite	PMMA	Rock salt	
Conductivity, Ω^{-1} m^{-1}	8.1×10^6	8.1×10^5	1.6×10^4	[2, 3]
Current density, A m^{-2}	1.1×10^{11}	1.6×10^{10}	1.7×10^{11} 6.1×10^{11}	[1–3]
Particle concentration in the channel, m^{-3}	–	2.9×10^{28}	–	[4]
Temperature in the channel, K	–	2.0×10^4	$(0.3–0.5) \times 10^4$	[1, 3, 4]
Specific energy content of the channel, J m^{-3}	2.5×10^{11}	2.2×10^{10} 8.5×10^{10}	10^{10}	[2, 3, 5, 6]
Power density, W m^{-3}	1.2×10^{17}	10^{16}	$10^{17}–10^{19}$	[1, 4]
Pressure, Pa	–	3×10^9	$(3–15) \times 10^9$	[4, 7, 8]

We note that the discharge current amplitude is 10^3–10^5 A and the duration of energy deposition period is (10^{-6}–10^{-4}) s for the majority of EDT variants. In this case, the characteristic energy content is (10^3–10^4) J/m, the channel radius is (5×10^{-3}–2×10^{-2}) m, its length is (10^{-2}–1) m, the temperature of the substance in the channel is approximately 10^4 K, and the pressure in the spark channel is (4×10^8–5×10^9) Pa. The current density for these conditions is sufficiently small to compete with the magnetic pressure component, and it can be considered that the kinetic pressure component plays the key role and determines the dynamics of spark expansion. Such conclusion is made by the authors of the majority of works devoted to a study of the channel stage of the electric discharge in solid dielectrics.

4.1.1 Discharge Channel as an Electrical Load

The breakdown channel in the discharge circuit of the energy unit acts as an active load, and the process of energy release can be characterized by the following parameters: discharge current i, voltage drop on the breakdown channel U_{ch}, its active resistance R_{ch}, power N_{ch} developed in the channel, energy W_{ch} liberated in the channel by the given moment of time.

In this case, the relationships

$$R_{ch} = U_{ch} i^{-1}, \quad N_{ch} = U_{ch} i, \quad W_{ch} = \int_0^t N_{ch} dt \tag{4.1}$$

are valid. High electric strength of most solid dielectrics (including strong rocks) in combination with relatively low energy required for effective destruction of dielectrics make the deeply oscillating discharge regime of the storage capacitor in the discharge circuit comprising the spark channel in the solid dielectric as a unique useful load [9]. The discharge current in this situation can a priori be predicted, because it is limited mainly by the internal impedance of the generator, and the active electric resistance R_{ch} of the spark channel is the base quantity for construction of calculation schemes, development of devices of ED technology, and search for other electrical characteristics of the channel.

Theoretical prediction of the function $R_{ch}(t)$ is based on a priori notions about the kinetics of plasma of the current column and dynamics of changes of its radius. Indeed, in the first approximation the breakdown channel in solid dielectrics has cylindrical shape, and for the cylinder

$$R_{ch} = l_{ch} / \pi r_{ch}^2 \langle \sigma \rangle \tag{4.2}$$

where r_{ch} and l_{ch} are the radius and length of the breakdown channel, $\langle \sigma \rangle$ is the electrical conductivity averaged over the cross section of the channel plasma.

Toepler [9], analyzing the gas gap in the glow discharge regime, derived and then generalized to the spark the expression

$$R_{ch} = K_0 l_{ch} \left(\int_0^t i(t)dt \right) \tag{4.3}$$

where $K_0 = (1\text{–}2) \times 10^6\,\Omega\,\mathrm{K\,m^{-1}}$.

Reflecting a larger contribution of the current compared to the time to the active resistance of an underwater spark, the correction coefficient $\exp(\alpha t)$ ($\alpha = \text{const}$) was added to the right-hand side of Eq. (4.3). In this case, satisfactory agreement with experiment was obtained for discharges close to critical ones when

$$R_{ch} \approx 2\sqrt{L/C}.$$

Under assumption that for a spark in gases the total deposited energy is equal to the internal energy of channel products, Rompe and Weizel [10] obtained

$$R_{ch} = l_{ch} \left(2\alpha_1 \int_0^t i^2(t)dt \right)^{-0{,}5}, \tag{4.4}$$

where $\alpha_1 = \text{const}$ depending on the gas sort.

Indeed, taking into account that $\sigma = n_{in} k_{in} e$, where n_{in} and k_{in} are the electron concentration and mobility,$i = \pi r_{ch}^2 n_{in} k_{in} E e$ and $R_{ch} i = l_{ch} E$, from Eq. (4.2) we have

$$R_{ch} = l_{ch} \left(\pi r_{ch}^2 n_{in} k_{in} e \right)^{-1}. \tag{4.5}$$

On the other hand, reducing the expression for the internal energy W_{in} to the form $W_{in} = \omega V_{ch}$, where $V_{ch} = \pi r_{ch}^2 l_{ch}$, we obtain

$$\begin{aligned} W_{in} &= \frac{2}{3}\pi r_{ch}^2 l_{ch} n k T + \pi r_{ch}^2 l_{ch} n_{in} e V_i + \pi r_{ch}^2 l_{ch} \left(n_a + \frac{D}{m_a} \right) \\ &= r_{ch}^2 l_{ch} n_{in} \left[\frac{3}{2} k T \left(1 + \frac{1}{x_i} \right) + e V_i + \frac{D}{x_i m_a} \right], \end{aligned} \tag{4.6}$$

where m_a is the number of atoms in the molecule and x_i is the ionization degree.

Comparing Eqs. (4.5) and (4.6), we have

$$R_{ch} = \frac{l_{ch}^2}{W_{in}} \left[\frac{3}{2} k T \left(1 + \frac{1}{x_i} \right) + e V_i + \frac{D}{m_a} \right] (k_{in} e)^{-1}.$$

If we set $\alpha_1 = k_{\mathrm{in}} e \left[\frac{3}{2} k T \left(1 + \frac{1}{x_i} \right) + e V_i + \frac{D}{m_a} \right]^{-1}$, then

$$\alpha_1 = k_{in} e \left[\frac{3}{2} kT \left(1 + \frac{1}{x_i} \right) + eV_i + \frac{D}{m_a} \right]^{-1},$$

$$R_{ch} = \frac{1}{\alpha_1} \frac{l_{ch}^2}{W_{in}}. \quad (4.7)$$

With allowance for the basic assumption

$$i^2 R_{ch} dt = dW_{in} \quad (4.8)$$

after integration of Eq. (4.8) with allowance for Eq. (4.7), we obtain Eq. (4.4).

The plasma temperature in the breakdown channel of solid dielectrics during one-two discharge current oscillation half-cycles is quasi-constant, and there are grounds to believe that α_1 is also roughly constant. Indeed, in [11] Eq. (4.4) was generalized to an underwater spark and in a wide range of variations of the discharge circuit parameters it was demonstrated that $(2\alpha)^{-1}$ changed rather weakly—from 0.22×10^5 to 0.28×10^5 V^2 s m^{-2}.

In addition to the approximations indicated above, a number of other approximations comprising $i(t)$, t, and one-three constants depending on the discharge regime, initial conditions in the discharge gap, and other factors can be found in the literature.

Insufficient degree of elaboration of theoretical knowledge on the dense low-temperature plasma kinetics in combination with almost insurmountable difficulties of theoretical prediction of $r_{ch}(t)$ without application of the energy balance equation including the function $R_{ch}(t)$ forces one to develop representations about $R_{ch}(t)$ on the empirical and semi-empirical bases. Such method is sufficiently efficient, because the procedure of experimental determination of $R_{ch}(t)$ (with error acceptable for practical application of results) is simple and easily reproduced.

Objects of the EDT are exclusively diverse by their properties regarding the composition of the plasma formed due to solid dielectric breakdown and resistance of the substance to expansion of the breakdown channel. In addition, for the preset interelectrode gap, the length of the breakdown channel, subjected to significant static fluctuations from experiment to experiment, also depends on the dielectric material (a crystal, an amorphous body, presence of inclusions, and so on). Around the channel of breakdown, radial kinks are sometimes formed at an early stage, the cavities of which are closed in a certain time with the channel; while the plasma will penetrate into these cavities.

Considering so numerous factors determining the active resistance of the breakdown channel in a solid dielectric, it is difficult to hope for a universal approach to exact a priori determination of $R_{ch}(i, t)$ for a wide range of variations of the dielectric properties, interelectrode gaps, and discharge circuit parameters.

In this connection, one of the authors of this monograph (N. Zinov'ev) performed measurements of $R_{ch}(t)$ during the first current oscillation. The results were systematized by approximation $R_{ch} = f(t)$ or $R_{ch} = f(i, t)$ with simple expressions containing only one dimensional constant whose value was related with the physicomechanical properties of the dielectric.

In the first series of experiments, the spark channel arose during breakdown of flat samples imbedded into transformer oil in the *rod-rod* system of electrodes placed at the opposite sides of the sample.

Five materials were chosen for investigation, including KCl single crystal, organic glass, and rocks: sandstone (Baleiskii Graben in the Chita Region), granite (Altai Mountains), and microquartzite (Leninogorskii deposit). A VIG was used as a source of high voltage impulses. The discharge capacitance C, inductance L, and impulse amplitude U_0 varied as given in Table 4.2. For each combination of experimental conditions, 8–10 samples with thicknesses from 5 to 50 mm were used. The breakdown was initiated at the impulse front with duration of 0.2 μs. Errors in measuring of the parameters were 0.5% for the sample thickness, 7% for the voltage impulse amplitude U_{ch}, 8% for the current i, 7.5% for time intervals, and 15% for R_{ch}. Results of experiments were processed to test the dependence $R_{ch}(t)$ given by Eq. (4.4).

In the derivation of Eq. (4.4) in [10] it was assumed that all deposited energy was spent on the increase of the internal energy, i.e., Equation (4.8) is valid and it will remain valid to within the constant A if we assume that in addition to the internal energy, other components are also important, but their contributions do not change with time

$$R_{ch} = \frac{Al_{ch}}{\sqrt{\int_0^t i^2 dt}} \quad A = f(t, \text{material}). \tag{4.9}$$

It is suffice to consider $\eta = \frac{A}{W_{ch}} \neq f(t)$; then instead of Eq. (4.8), we have

$$(1 - \eta) dW_{in} = i^2 R_{ch} dt, \tag{4.10}$$

$$A = \frac{1}{\sqrt{(1 - \eta)^{2a}}} = \text{const} \tag{4.11}$$

To test experimentally the applicability of Eq. (4.9), it is suffice to demonstrate that the dependence $A = f(t)$ can be neglected; in this case, the influence of the interelectrode gap $A = f(S)$ can be considered or ignored if it is small.

As an example, Fig. 4.1 shows the dependences $A = f(t)$ for the first discharge current oscillation for three materials and different regimes of energy deposition. The errors in determining A values were ~20%, within the same limits changes the error in determining $A = f(t, \tau)$, where $\tau \approx \pi\sqrt{LC}$ is the duration of current oscillation. The least spreads were observed for the breakdown of the KCl single crystal. This was due to the stability of the properties from sample to sample and strict orientation of the channel along the |100| crystallographic direction; this provided the constant channel length. The greatest spreads were observed for rocks, especially for the microquartzite whose samples contained foreign impurities.

Table 4.2 Values of the spark constant A_{av} and its confidence interval ε′, V $s^{0.5}$ m^{-1}

Serial number	Discharge conditions	KCl	Plexiglass	Sandstone	Granite	Quartzite
1	L=4.7 μH C=0.0167 μF U_0 = 210 kV	251	280	528	680	618
2	L=14.7 μH C=0.0417 μF U_0 = 210 kV	214	346	425	630	746
3	L=49.2 μH C=0.0417 μF U_0 = 210 kV	216	408	609	715	973
4	L=10.3 μH C=0.0417 μF U_0 = 210 kV	234	279	568	642	941
5	L=15.7 μH C=0.0583 μF U_0 = 180 kV	246	–	–	–	–
6	L=15.7 μH C=0.0583 μF U_0 = 210 kV	–	404	493	648	763
7	A_{av}	232	343	525	663	809
8	ε′	33	88	96	48	205

The assumption about the change of the channel length with respect to the increasing interelectrode gap as a reason for the change of A attendant to variations of S was not confirmed for the KCl single crystal for which the channel length strictly corresponded to the interelectrode gap. It is more reasonably to associate A=$f(S)$ with the outflow of the products of the breakdown channel into the liquid surrounding the solid dielectric sample. For small interelectrode gaps, the loss of the current carriers caused A to increase. This conclusion was confirmed by the experiment with the camouflage-type discharge in KCl samples. (Measures on prevention of plasma outflow from the channel mouths into the environment by means of electrode crimping by epoxy compound were undertaken.) As can be seen from Fig. 4.2, A_{av} = 220 V $s^{0.5}$ m^{-1} for S=5 mm, which corresponds to the A_{av} value for S = 30–50 mm, when the

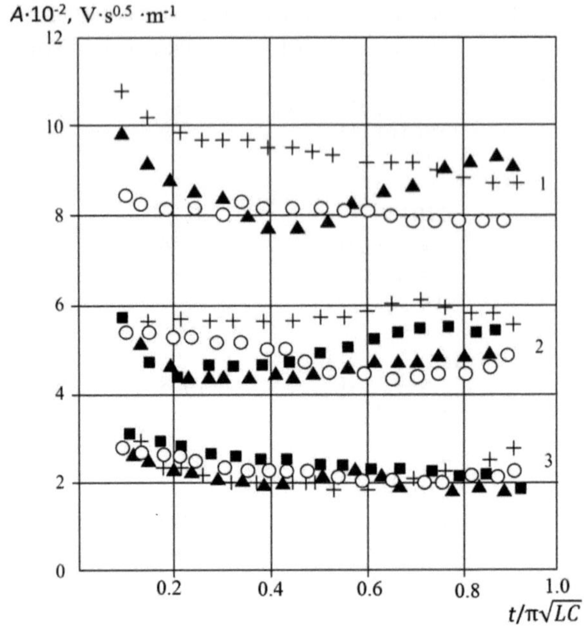

Fig. 4.1 Dependence of the spark constant A on time t for microquartzite (curve 1), sandstone (curve 2), and KCl (curve 3). Here $\tau = 1.8$ (black square), 2.4 (white circle), 4.5 (plus sign), and 6.5 μs (black up-pointing triangle)

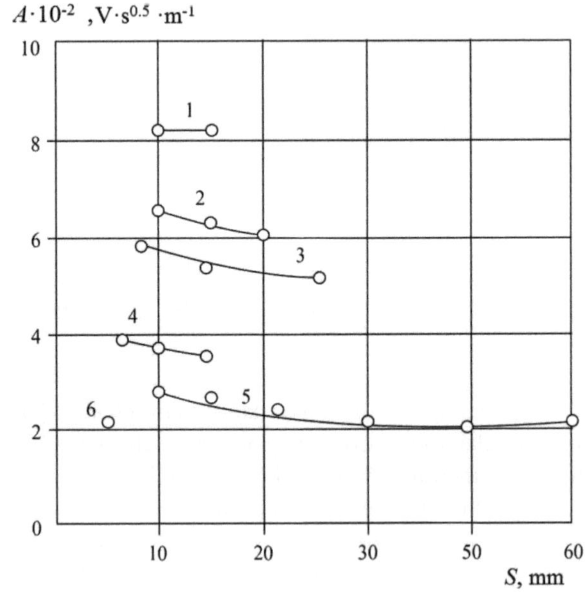

Fig. 4.2 Dependence of the spark constant A on the interelectrode gap length l_e for microquartzite (curve 1), granite (curve 2), sandstone (curve 3), plexiglass (curve 4), KCl (curve 5), and KCl with camouflage discharge (curve 6)

effect of outflow of the channel products on the dynamics of energy deposition is no longer significant.

To characterize the ability of the material to convert the electric field energy into the internal energy of plasma in the breakdown channel, it is expedient to use the A values in the segment where $A \neq f(S)$. The average values A_{av} of the spark constant of the indicated materials and their confidence intervals are given in Table 4.2 for $S = 10$ mm.

When solving the problems of the transient process in the circuits comprising only the spark channel as a load, formula (4.9) gives $R_{ch} = \infty$ at $t = 0$, which makes the initial conditions uncertain. In this case, the approximation

$$R_{ch} = \frac{A l_{ch}}{\Delta + \sqrt{\int_0^t i^2 dt}} \tag{4.12}$$

is expedient, where $\Delta \approx \pi A r_0^2 \sigma_0$.

According to the literature sources, the initial radius r_0 of the channel and the initial electrical conductivity σ_0 for alkali-halide crystals can be set equal to 1 μm and $\sigma_0 \sim 1.6$–$2.6\ \Omega^{-1}\ \text{m}^{-1}$, respectively.

To generalize approximations (4.9) and (4.12) to solid materials that have not been investigated yet, it is expedient to relate the spark constant to the physicomechanical properties of the dielectric.

High degree of correlation of the active resistance of the spark channel in solid dielectrics with their mechanical and acoustic properties is observed. The last is predictable if we consider that the acoustic rigidity is $\lambda_0 = p_0 c_0 = P/u$, where u is the mass velocity of the medium under the action of force (pressure) P; p_0 and c_0 are the density of the medium and the sound speed in it.

In the spark channel

$$u \sim \frac{dr_{ch}}{dt} \sim \frac{r_{ch}}{t} \sim \frac{P(\lambda_0)}{\lambda_0}.$$

On the other hand, the relationship $R_{ch} \sim \frac{1}{r_{ch}^2 \langle\sigma\rangle}$ holds true.

The electrical conductivity of a substance in the breakdown channel depends on the temperature, pressure, and elemental composition of plasma in the spark channel. The plasma temperature is quasi-constant for widely varying regimes of energy deposition and types of dielectrics, whereas the pressure and elemental composition depend significantly on the properties of the dielectrics. At $T = 10^4$–2×10^4 K and $P \sim 10^7$–10^9 Pa, $<\sigma>$ depends weakly on the plasma composition, but rather noticeably on the pressure. In this case, it can be noted that for $P = 2 \times 10^8$ Pa and $T = 1.8 \times 10^4$ K (characteristic for technological application of impulse discharges in solid dielectrics), $<\sigma> \sim 1.5\ \Omega^{-1}\ \text{m}^{-1}$. The strongest deviation is observed for hydrogen plasma ($<\sigma> = 0.3\ \Omega^{-1}\ \text{m}^{-1}$); for plasma of some metals, including halide ones, $<\sigma> = 2.0$–$2.5\ \Omega^{-1}\ \text{m}^{-1}$. The characteristic feature of $\sigma(P)$ is a weak linear growth in the region $P = (10^5$–$5 \times 10^8)$ Pa and intensive nonlinear increase for $P > 5 \times 10^8$ Pa.

Taking into account the dependences $P=f(\lambda_0)$ and $\sigma=f(P)$, the dependence $\sigma=f(\lambda_0)$ arises, that is, a strong correlation between R_{ch} and λ_0 can be expected.

In view of the foregoing, a relationship between the spark constant and the acoustic rigidity λ_0 of the materials was found. To this end, the density of the examined dielectrics was measured by the standard pycnometric technique, and the sound speed was measured by means of an UZIS-LE ultrasonic speedometer. Results of measurements are given in Table 4.3. The characteristics for water were borrowed from reference literature. The dependence $A=f(\lambda_0)$ shown in Fig. 4.3 is approximated sufficiently well by the formula

$$A = 1000\left(1 - e^{-k_1\lambda_0}\right), \tag{4.13}$$

where A is in V $s^{0.5}$ m^{-1} and $k_1 = 10^{-7}$ kg^{-1} m^2 s^1.

Equation (4.13) is valid for the materials with acoustic rigidity $\lambda_0 = (1.5–14.5) \times 10^{6\,kg}$ m^{-2} s^{-1}.

Almost all EDT objects fall within this range. One of two reasons can explain the deviation of KCl from this general law in Fig. 4.3.

Table 4.3 Characteristics of materials

Dielectric	c_0, m s^{-1}	ρ_0, kg m^{-3}	$\lambda_0 \times 10^{-6}$, kg m^{-2} s^{-1}
Quartzite	5300	2750	14.57
Granite	4435	2770	11.40
Sandstone	2860	2660	7.60
Concrete M–200	2360	2200	5.19
Plexiglass (PMMA)	2760	1180	3.15
KCl	4070	1950	7.94
Water	1497	1000	1.50

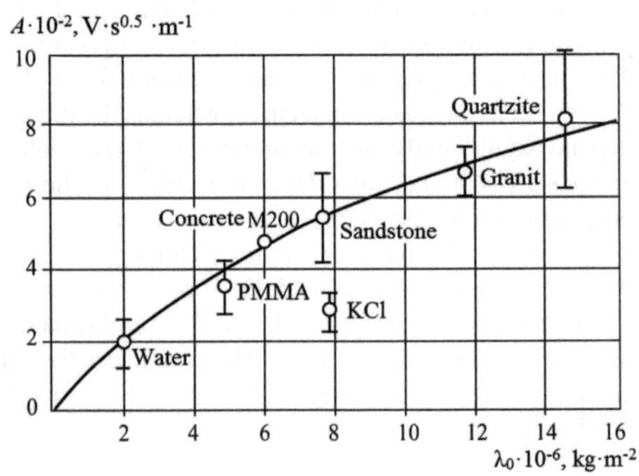

Fig. 4.3 Dependence of the spark constant A on the acoustic rigidity λ_0 of the medium

Table 4.4 Parameters of the discharge circuit

Regime	l_{ch}, m	U_0, kV	C_d, μF	L, μH	W_0, J	$Z_W = \sqrt{L/C_d}$, Ω
1	0.1	192	0.058	16	1075	16.6
2		300	0.058	16	2625	16.6
3		330	0.058	64	3180	33.2
4		360	0.058	16	3780	16.6
5		360	0.058	64	3780	33.2
6		360	0.058	96	3780	40.7
7		408	0.058	16	4855	16.6
8	0.2	390	0.058	830	4440	119.6
9		402	0.058	96	4710	40.7
10		420	0.058	16	5145	16.6

1. The spark constant was defined as a product of the square root of the current action integral by the active electric resistance of the channel normalized by the interelectrode gap rather than by the channel length as it should be.
2. The length of the breakdown channel significantly exceeds the interelectrode gap length due to large- and small-scale curvatures. Thus, in the case of ore breakdowns [1], the channel length reaches 1.5 of the interelectrode gap length. For KCl samples with |100| crystallographic direction along the gap, the channel length l_{ch} is equal to the gap length, that is, the true A value is determined for this material; for all other dielectrics, A is overestimated by a factor of l_{ch}/S (where l_{ch} is the channel length and S is the distance between the electrodes equal to the thickness of the sample).

It is also possible that deviation of the KCl point from the general law is partially caused by a stronger (compared with other materials) influence of the plasma electrical conductivity on the value of the resistance of the spark channel because of low ionization potential of the potassium atom.

The single-crystal structure is not characteristic for ores, rocks, building materials, and other EDT objects. The high content of alkaline metals is also not characteristic for them; therefore, the use of Eq. (4.13) is justified for aprioristic determination of A in most cases.

In the second series of experiments, the resistance of the spark channel in concretes is investigated for decimeter interelectrode gaps.

Experiments were performed with samples of concrete brand 200, 300, 400, 500, and 600. The VIG parameters were varied by changing the inductance L of the discharge circuit and the voltage U_0 in accordance with Table 4.4. The breakdown of samples of concrete with thicknesses of 10 and 20 cm was initiated in transformer oil for the tip-pane system of electrodes arranged on the opposite sides of the sample. To find R_{ch}, the discharge current and the voltage drop across the sample were recorded.

For each combination of experimental conditions, 7–10 samples were subjected to breakdowns. The spark constant A_{ch} was determined from Eq. (4.9) after substitution

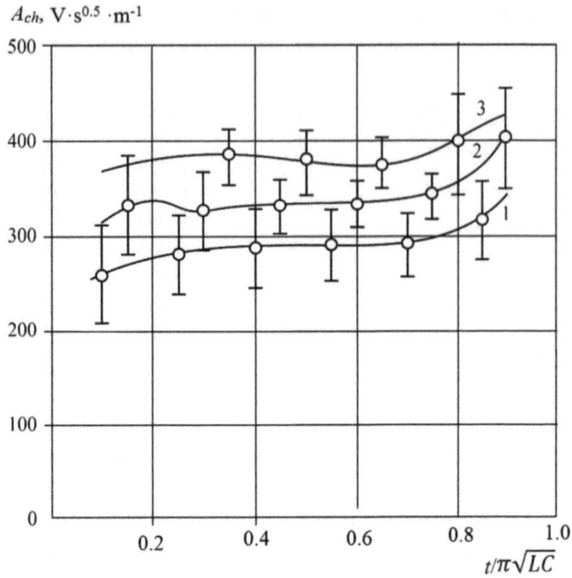

Fig. 4.4 Time dependence of the spark constant A for concrete brand 200 (curve 1), 400 (curve 2), and 500 (curve 3)

of experimental values $R_{ch}(t)$ and $i(t)$ in this equation. The length l_{ch} of the spark channel was set equal to the interelectrode gap length.

Figure 4.4 shows variations of the spark constant throughout the first current period for concretes brand 200, 400, and 600 and different regimes of energy release. As can be seen from the figure, A is quasi-constant for the most part of the half-period τ that confirms the applicability of Eq. (4.9) for a description of the spark resistance in concretes. The wide spread of the experimental data demonstrates the instability of the breakdown channel caused by extremely non-uniform structure of concrete and different compositions of the test samples even for one and the same brand of concrete. In calculations of the channel resistance, the average value A_{av} for the period (0.1–0.9)τ can be considered when the most part of energy is released:

$$A_{av} = \frac{1}{0.8\tau}\int_{0.1\tau}^{0.9\tau} A dt.$$

It seems likely that the increase of the spark constant and hence of the resistance for the period (0.9–1)τ is due to the decrease of the electrical conductivity of the channel owing to a decrease in the plasma cord temperature. The increase of the spark constant by the end of the half-period compared to its average value A_{av} has no significant effect on the energy deposition because of the small discharge current.

Investigations of the spark constant depending on the interelectrode distance for concrete brand 200 demonstrated that at $l_k l_{ch}$=0.1 and 0.2 m, the A_{av} value was equal to 290 and 279 V s$^{0.5}$ m^{-1}, respectively. Insignificant difference of the data obtained allows the conclusion to be made that in the examined period the influence of l_{ch} on A_{av} can be neglected.

Results of investigations of the spark constant of concrete of different brands are shown in Fig. 4.5.

Estimates of the acoustic rigidity of concrete have shown that λ_0 values increased from 8.7×10^6 to 1.1×10^7 kg m^2 s^{-1} when the brand of concrete changes from 200 to 600. Probably, this effect also explains weak dependence of the spark constant on the brand of concrete (Fig. 4.5). The recommended values A_{av} for calculations of the resistance of the spark channel in concrete are presented in Table 4.5.

Thus, based on the results of experimental research of samples of concrete, it is possible to conclude that a weak dependence of the spark constant on the acoustic rigidity is characteristic for concrete of different brands. Experiments have confirmed the applicability of Eq. (4.9) for the estimation of the resistance of the breakdown channel in concrete. This allows, using relations (4.9), the energy characteristics of the spark in concrete to be calculated and the parameters of material flow around the breakdown channel to be determined.

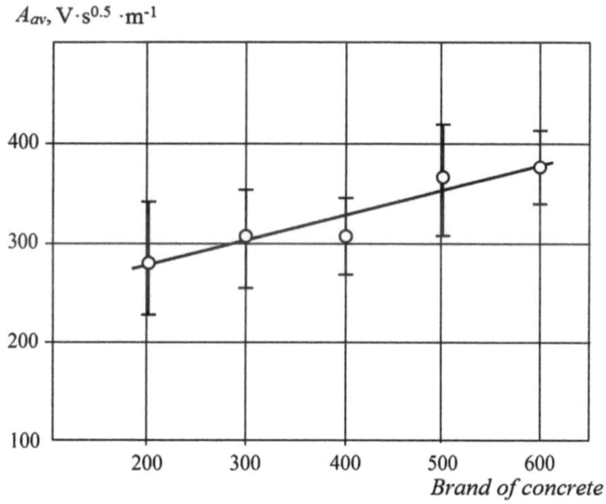

Fig. 4.5 Dependence of A_{av} on the brand of concrete

Table 4.5 Spark constant of concrete

Brand of concrete	200	300	400	500	600
A_{av}, V s$^{0.5}$ m^{-1}	290	305	325	350	375

4.1.2 Energy Characteristics of the Discharge Channel. Transient Processes

The efficiency of the destruction action of the impulse electric discharge is determined by the maximal pressure P_{ch} developed in the breakdown channel, impulse S_{pw} of the pressure wave, energy W_{cw} of the wave, and also their variation with time. In turn, for the preset electric energy released in the spark, P_{ch}, S_{pw}, and W_{cw} values are determined by the law of energy release in the current cord plasma.

By the energy characteristics of the electric discharge in solid dielectrics, the discharge current $i(t)$, the voltage drop $U(t)$, the resistance $R(t)$ of the channel, and the power $N(t)$ developed in the channel are understood.

The notion on the influence of the individual discharge circuit parameters on these characteristics of the simplest circuit with lumped parameters can be obtained by solving the problem on transient processes in the *R–L–C* circuit. The transient process in this case is described by the differential equation

$$L\frac{di}{dt} + R_{ch}(i,t)\cdot i + \frac{1}{C}\int_{0}^{t} idt = U_0, \quad i(0) = 0. \tag{4.14}$$

No analytical solution can be obtained for the transient process with the dependence R_k given by Eq. (4.9), whereas numerical solution of system of Eqs. (4.9) and (4.14) by simple exhaustive method of discharge circuits will yield a large number of variants to analyze them all to reveal regularities is quite difficult and sometimes even impossible. The variable substitutions

$$x = \frac{t}{\sqrt{LC}}, \quad y = \frac{i}{U_0}\sqrt{\frac{L}{C}}, \quad m = R_{ch}\sqrt{\frac{C}{L}} \tag{4.15}$$

Reduce Eqs. (4.9) and (4.14) to the system of dimensionless equations of the form

$$\frac{dy}{dx} + my + \int_{0}^{x} ydx = 1, \quad y(0) = 0, \tag{4.16}$$

$$m = \frac{Al_{ch}}{U_0\sqrt[4]{LC}} \cdot \frac{1}{\sqrt{\int_0^x y^2 dx}}, \tag{4.17}$$

In system (4.16) and (4.17), all variety of variable discharge circuit parameters is reduced to their combination representing the dimensionless complex, being the similarity criterion (CS_{br}) of the electrical characteristics of the spark channel:

$$CS = \frac{Al_{ch}}{U_0\sqrt[4]{LC}}, \tag{4.18}$$

This means that two spark discharges with different L, C, U_0, A, and l_{ch} for $CS_1 = CS_2$ will be described by identical dimensionless functions of the current $y = f(\tau, (CS_{br})$, resistance $m = f(\tau, (CS_{br})$, power $\lambda = f(\tau, (CS_{br})$, and deposited energy $\hat{w} = f(\tau, (CS_{br})$. To test the similarity, we took advantage of the results obtained by different authors at the HVRI. In these works, it was impossible to determine the value of the spark constant; therefore, the reduced criterion

$$CS_{br} = \frac{l_{ch}}{U_0 \sqrt[4]{LC}} \tag{4.19}$$

was used valid for each individual material.

Figures 4.6 and 4.7 show the dimensionless functions of power and energy from the results of experimental works for sandstone

$$\lambda_m = \frac{2N\sqrt{LC}}{CU_0} f(CS_{br}), \tag{4.20}$$

$$\hat{w}_m = \frac{2W}{CU_0} f(CS_{br}). \tag{4.21}$$

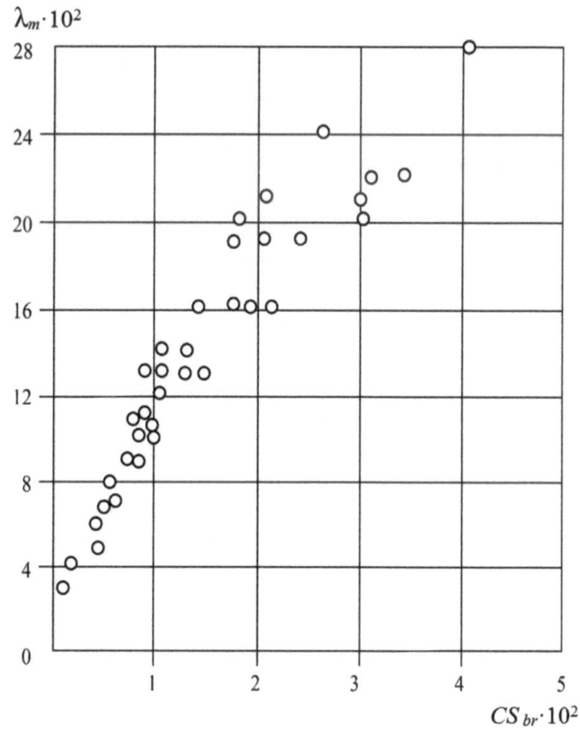

Fig. 4.6 Normalized power as a function of the reduced criterion CS_{br}

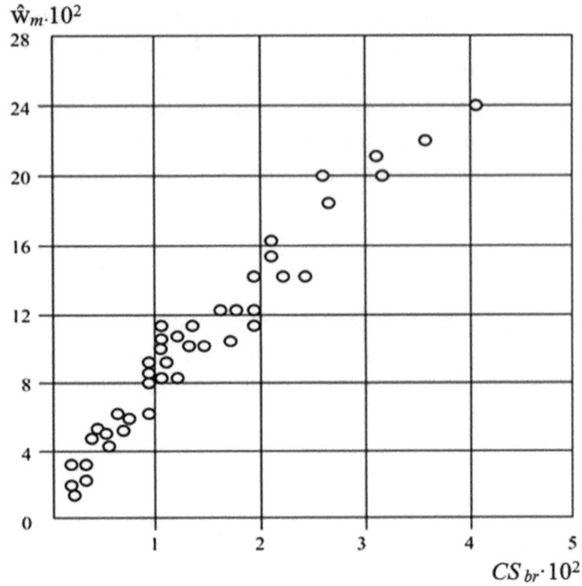

Fig. 4.7 Normalized energy as a function of the reduced criterion CS_{br} at the moment of current maximum

The subscript m in the figure indicates that the values of the functions were determined at the moment of current maximum.

Figures 4.8 and 4.9 show the dependences of the dimensionless power and resistance functions on the criterion CS_{br} that, along with the energy release regime, considers the influence of the material properties on the energy characteristics of the discharge.

To construct the above-indicated dependences in the range falling outside of the limits of the possibilities of our experimental facility, system (4.16) and (4.17) in which the variable parameter was $CS_{br} \times 10^2$ was solved by a numerical method for the first period of the discharge current for the initial conditions $\tau = 0$, $y = 0$, and $\dot{y} = 0$.

The calculated functions are presented in the form of plots varying with time at fixed CS_{br} values, Fig. 4.10.

For the spark discharge in condensed media, the active resistance is essentially nonlinear element; its value changes by several orders of magnitude (from several tens of kiloohms to several fractions of the ohm) during small time periods. The discharge character during the same time changes from aperiodic to deeply oscillatory. In calculations for small values $CS_{br} < 0.05$, the stage of the aperiodic discharge can be neglected, since $t_1 \ll \sqrt{LC}$, where t_1 is time during which the condition of the discharge aperiodicity is satisfied.

Thus, practically during all time of the transient process the discharge is deeply oscillatory, sinusoidal in character; moreover, the resistance of the spark channel has almost no effect on the oscillation amplitude and period. It should be noted that such

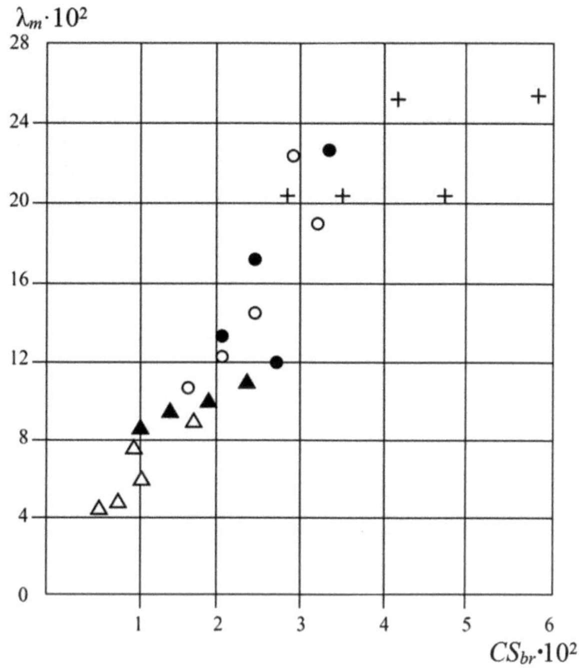

Fig. 4.8 Normalized values of the power at the moment of current maximum. Here white triangle are for KCl, plus sign are for quartzite, white circle are for granite, black circle are for sandstone, and black up-pointing triangle are for PMMA

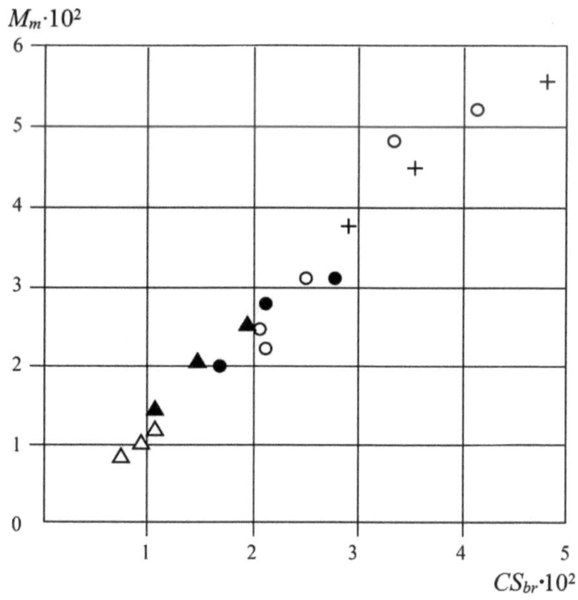

Fig. 4.9 Normalized values of the resistance at the moment of current maximum. Here white triangle are for KCl, plus sign are for quartzite, ○ are for granite, black circle are for sandstone, and black up-pointing triangle are for PMMA

situation is typical for disintegration and is a consequence of high electric strength of solid dielectrics.

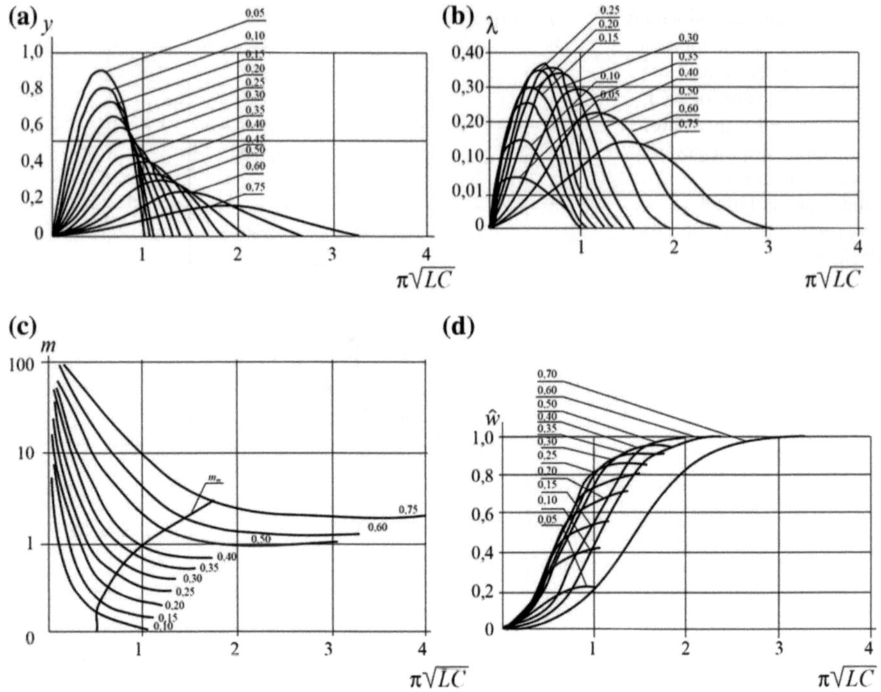

Fig. 4.10 Dimensionless functions of the discharge current (**a**), power (**b**), discharge channel resistance (**c**), and energy released in the channel (**d**)

For the discharge current in the first approximation the formula

$$i = U_0\sqrt{\frac{C}{L}} \cdot \sin \omega_1 t, \tag{4.22}$$

holds true, where $\omega_1 = \pi\tau_1^{-1} = (LC)^{-0.5}$. It allows analytical expressions to be derived for the energy characteristics of the discharge channel:

$$R_{ch} = \frac{Al_{ch}}{\sqrt{\int_0^t i^2 dt}} = Al_{ch}U_0^{-1}C^{-3/4}L^{1/4}\left[\frac{t}{\sqrt{LC}} - \frac{1}{2}\sin\left(2\frac{t}{\sqrt{LC}}\right)\right]^{-1/2}, \tag{4.23}$$

$$N = i^2 R_{ch} = Al_{ch}U_0^{-1}C^{1/4}L^{-3/4}\left[\frac{t}{\sqrt{LC}} - \frac{1}{2}\sin\left(2\frac{t}{\sqrt{LC}}\right)\right]^{-1/2} \cdot \left(\sin\frac{t}{\sqrt{LC}}\right)^2, \tag{4.24}$$

$$W_{ch} = \int_0^t i^2 R_{ch} dt = 2Al_{ch}U_0 C^{3/4}L^{-1/4}\left[\frac{t}{\sqrt{LC}} - \frac{1}{2}\sin\left(2\frac{t}{\sqrt{LC}}\right)\right]^{1/2}. \tag{4.25}$$

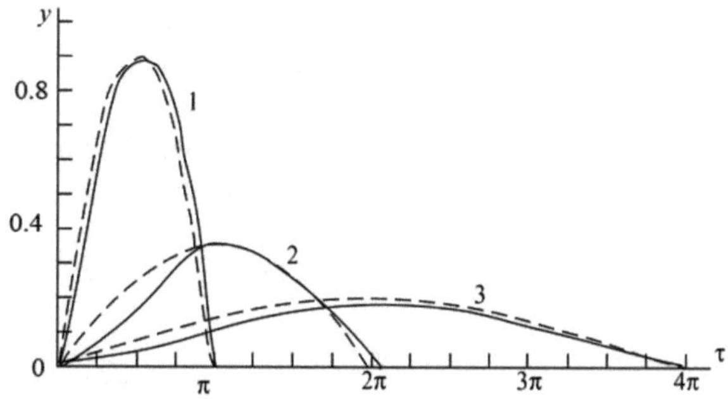

Fig. 4.11 Dependence of the normalized current y on the normalized time τ (during the first oscillation period) for $CS_{br} = 0.05$ (curve 1), 0.5 (curve 2), 0.75 (curve 3). The continuous curves are for the results of calculations, and the dashed curves are for the approximation curves

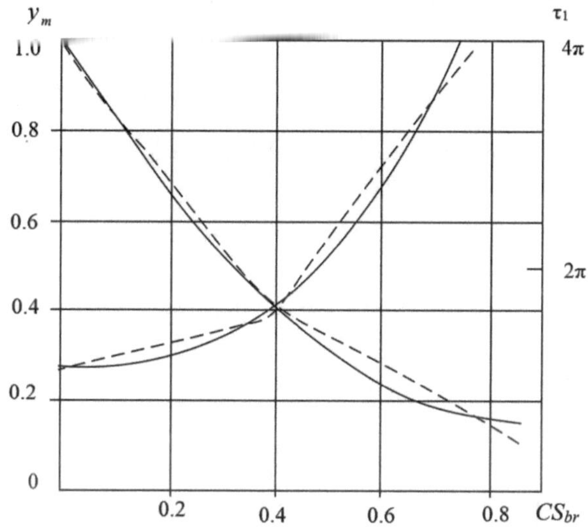

Fig. 4.12 Dependence of the normalized current amplitude y_m and of the duration τ_1 of its of the first half-period oscillations on the similarity criterion CS_{br}. Here the solid curves are for the results of calculations and the dashed curves are for the approximation curves

For $CS_{br}>0.75$, it is necessary to consider corrections taking into account the effect of the spark channel resistance on the discharge current amplitude and oscillation period. Figure 4.11 shows the dependences of the discharge current at three CS_{br} values.

In the first quarter of the oscillation period of the discharge current for $CS_{br}=0.5$, the effect of the spark resistance on the discharge current waveform is noticeable. The discharge is aperiodic in character and goes into oscillatory one with decreasing resistance.

For $CS_{br}>0.75$, the discharge is aperiodic in character during the entire time of the condenser discharge. However, the waveform of the current $y(x)$ can also be approximated fairly well by a sinusoidal function.

To determine the amplitude y_m and the half-period of the function being approximated, plots of the dependences $y_m=f_1(CS_{br})$ and $\tau_1 = f_2(CS_{br})$ are drawn in Fig. 4.12.

For $0<CS_{br}\leq 0.75$, we take advantage of a piecewise-linear approximation:

$$y_m = 1.5(0.67 - CS),\ \tau_1 = \pi(1 + 1.35CS) \qquad \text{by } 0.0 < CS_{br} \le 0.40;$$
$$y_m = 0.67(1.0 - CS_{br}),\ \tau_1 = \pi(7.021CS_{br} - 1.27) \text{ by } 0.40 < CS_{br} \le 0.75.$$

shown in Fig. 4.12 by the dashed curves. Taking into account the foregoing, the discharge current can be written as follows:

$$i = 1.5(0.67 - CS)V_0\sqrt{\frac{C}{L}}\sin\left(\frac{t}{\sqrt{LC}} + 1.35\pi CS_{br}\right), \quad \text{by } 0.0 < CS_{br} \le 0.40; \tag{4.26}$$

$$i = 0.67(1 - CS)U_0\sqrt{\frac{C}{L}}\sin\left[8.1\left(\pi CS_{br} - 0.21\frac{t}{\sqrt{LC}}\right)\right], \quad \text{by } 0.40 < CS_{br} \le 0.75. \tag{4.27}$$

Then the energy functions of the spark channel can be described by expressions analogous to formulas (4.23–4.25)

$$i = I_m \sin(\omega t + D) \tag{4.28}$$

$$R_{ch} = \frac{Al}{\sqrt{\int_0^t i^2 dt}} = 2Al_{ch}I_m^{-1}\omega^{0.5}[2\omega t - \sin 2(\omega t + D)]^{-0.5}, \tag{4.29}$$

$$N = i^2 R_{ch} = 2Al_{ch}I_m\omega^{0.5}[2\omega t - \sin 2(\omega t + D)]^{-0.5}[\sin(\omega t + D)]^2 \tag{4.30}$$

$$W_{ch} = \int_0^t i^2 R_{ch} dt = Al_{ch}I_m\omega^{-0.5}[2\omega t - \sin 2(\omega t + D)]^{0.5}, \tag{4.31}$$

where

$$\begin{cases} I_m = 1.5(0.67 - CS)U_0\sqrt{\frac{L}{C}},\ \omega = \frac{1}{\sqrt{LC}},\ D = 1.35\pi CS_{br}, \text{ by } 0.0 < CS_{br} \le 0.40, \\ I_m = 0.67(1 - CS)U_0\sqrt{\frac{L}{C}},\ \omega = \frac{-1.7}{\sqrt{LC}},\ D = 0.81\pi CS_{br}, \quad \text{by } 0.40 < CS_{br} \le 0.75. \end{cases}$$

Formulas (4.28–4.31) can be represented by the expressions of the form:

$$i = m_i f_i, \quad R_{ch} = m_R f_R, \quad N = m_N f_N, \quad W = m_W f_W,$$

where m_R, m_N, and m_W are the scale factors and f_R, f_N, and f_W are functions describing the time dependence of the resistance, power, and energy, shown in Fig. 4.13.

The efficiency of destruction owing to the solid breakdown by high-voltage impulses is related with the parameters of the shock waves generated in the discharge channel in the stage of its expansion. The pressure and the velocity of the shock wave front are rigidly connected with rate of power increase. If we replace the actual waveform of the power impulse by a triangular one, than the power amplitude N_{max} on the active element by the time t_{max} from the moment of the gap breakdown can be written as

$$\frac{dN}{dt} = \frac{N_{max}}{t_{max}}. \tag{4.32}$$

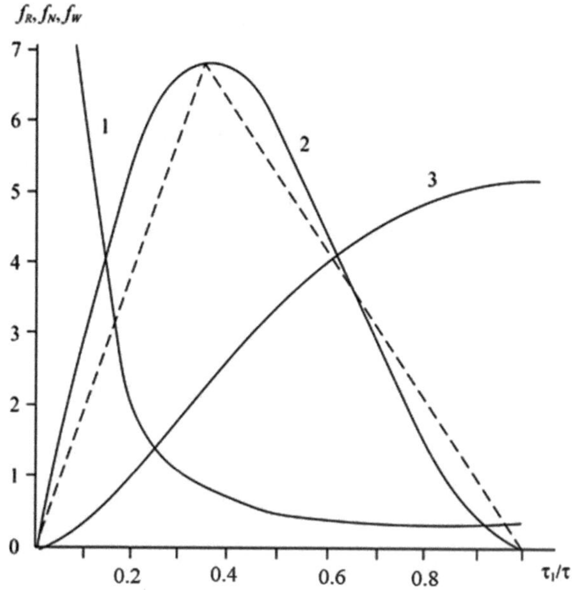

Fig. 4.13 Dimensionless functions describing the time dependence of the resistance f_R (curve 1), power f_N (curve 2), and energy f_W (curve 3). Here the solid curves are for the results of calculations, and the dashed curve shows the approximation curve for f_N

Thus, changing the power in the spark channel and the rate of its increase, it is possible to vary in wide limits the configuration of the mechanical stress field in the process of solid destruction.

4.1.3 *Energy Balance of the Channel Stage of the Discharge*

By analogy with the theory of electric discharge in liquids, the energy balance of the electric discharge in a solid dielectric can also be represented as follows:

$$W_{in} + A_{ch} + W_{rad} + G + W_{outfl} = W_{ch}. \tag{4.33}$$

Here W_{in} is the internal energy of the destruction products in the channel, A_{ch} is the work done by the discharge channel against the environment, W_{rad} is the energy loss on light emission, G is the heat loss on the heat conductivity, W_{outfl} is the loss with the substance that flows out through the channel mouth into the environment, and W_{ch} is the energy released in the channel.

Let us consider the terms of Eq. (4.33).

The spectrum of radiation of the discharge channel, both for the liquid and solid is close to the spectrum of the absolutely black body. Therefore, the radiated energy can be described by the Stefan–Boltzmann law:

$$W_{rad} = \sigma \int_{0}^{t} S_{ch}(t) T^4 dt, \quad (4.34)$$

where σ is the Stefan–Boltzmann constant, $S_{ch}(t)$ is the surface area of the discharge channel, and T is the temperature of the discharge channel.

Estimations by Eq. (4.34) demonstrate that energy of (0.3–0.9) J is emitted from 1 cm^2 of the discharge channel surface at $T = 15{,}000$–$20{,}000$ K for 10^{-6} s. According to the Wien law, the wavelength at which the energy of emission of absolutely black body is maximal is

$$\lambda_{max} \approx 0.2899 \cdot T^{-1}, \text{cm}$$

and is in the ultraviolet region of the spectrum

$$\lambda_{max} \sim 1.93 \times 10^{-5}\,\text{cm}(\text{at}\,T = 15,000\,\text{K}),$$
$$\lambda_{max} \sim 1.45 \times 10^{-5}\,\text{cm}(\text{at}\,T = 20,000\,\text{K}).$$

Almost all condensed media strongly absorb light radiation in this region. Thus, for example, at 1500 Å the absorption coefficient in water is about 10^4 cm^{-1}, and concrete is opaque to light radiation. In this regard, the energy losses on light radiation in water and the more so in concrete can be neglected.

Except photons, the energy (heat) is also carried away through the channel wall by atoms (or ions) and electrons via heat conductivity. In [12, 13] these losses were estimated for the electric discharge in water. The amount of the energy stored in a thin transient layer is expressed as

$$G = \omega\, S_{ch} \delta,$$

where ω is the energy of unit volume; S_{ch} is the surface area of the channel; δ is the thickness of the transient layer, $\delta = 2\sqrt{\chi^t}$; and χ is the coefficient of thermal diffusivity.

The discharge circuit in the experiments on the discharge in water had the following parameters: the capacitance of the energy storage device was $C = 1.5 \times 10^{-4}$ F, the inductance of the circuit was $L = 1.5 \times 10^{-6}$ H, the initial voltage of the energy storage device was $U_0 = 6$ kV, the distance between the electrodes was 7 cm, and the total amount of energy that escaped from the channel as heat was $G \sim 100$ J. This makes approximately 3% of the energy stored in the capacitor.

The order of magnitude of losses G calculated in a similar way for the electric discharge in solids was the same as for the discharge in water; they were several percent. In regimes of energy release in EDT installations, the temperature fields in the solid caused by the heat conductivity do not produce noticeable thermal stresses and do not influence cracking.

The outflow of hot substance from the channel mouths into the environment in the process of channel expansion (until the moment of closure of cracks with

Table 4.6 Parameters of plexiglas sample destruction for the indicated breakdown conditions (samples 75 × 75 × 22 mm, discharge gap of 8 mm, and generator energy of 200 J)

Measured parameter	Breakdown conditions		
	Both electrodes are embedded	One electrode is embedded	Electrodes are not embedded
Area of cracking polygon, mm^2	1052	615	330
Total length of visible cracks, mm	228	81	59
Area of visible cracks, mm^2	2552	836	339
Maximal crack length, mm	21	14	8
Duration of the cracking phase, μs	68	56	35

the channel column) during electric breakdown of solids causes the pressure in the channel to reduce together with efficiency of electric discharge. To estimate the effect of measures on prevention of plasma outflow, the electrodes were embedded into an epoxy compound. The destruction intensity was considered for three cases:

(1) plasma outflows freely,
(2) one electrode is embedded,
(3) both electrodes are embedded.

The energy of the vapor-gas void was calculated for three void shapes it acquired according to the hydrodynamics equations under various breakdown conditions: spherical and elongated and compressed ellipsoids of rotation.

It was demonstrated that during the period of crack growth, up to (30–50) % of the discharge channel energy is transferred with the substance into the vapor-gas void and the artificial delay of the plasma outflow increases the time of crack growth and the destruction volume, Table 4.6.

In general, for cylindrical symmetry of the channel of outflow through the mouth can be taken into account as follows:

$$W_{outfl} = 2\int_0^t \pi r_{ch}^2 \left(pu^2/2 + P_{ch}/\gamma_e - 1\right) \cdot u_z dt,$$

where r_{ch} is the channel radius; p and u are the density and mass velocity of the substance, outflowing into the environment through the cross section of the channel mouth; u_z is the projection of u onto the channel axis; and γ_e is the effective adiabatic index.

Typically, $pu^2/2 \ll P_{ch}/(\gamma_e - 1)$ in the stage of energy release, and then

$$W_{outfl} = \frac{2}{l_{ch}} \int_0^t V_{ch} \cdot P_{ch} \cdot (\gamma_e - 1)^{-1} u_z dt, \tag{4.35}$$

where l_{ch} is the channel length.

By a comparison of the camouflage discharge and the discharge with free plasma outflow in KCl for energy deposition period $t \sim (1\text{–}10) \times 10^{-6}$ s and other conditions remaining the same, it was experimentally established that the outflow process affects significantly the efficiency of the electric discharge provided that

$$l_{ch}/d_{ch} < 10, \tag{4.36}$$

where l_{ch} and d_{ch} are the length and the diameter of the channel.

In a number of practical cases (destruction of oversized samples, drilling of large-diameter wells, and destruction of reinforced concrete products to extraction the armature), the condition opposite to (4.36) is satisfied. In this situation, the effect of outflow in the phase of active energy release can be neglected, and according to the foregoing, the energy balance in the stage of energy release for the electric discharge in solid dielectrics can be simplified down to the expression:

$$W_{in} + A_{ch} = W_{ch}. \tag{4.37}$$

Certainly, unloading of the discharge channel should be considered in the technology of material disintegration using the scheme with one free surface, especially for short discharge gaps when unloading is observed not only through the discharge channel mouth, but also through cracks extended to the surface.

Fast electric energy release caused strong warming of the dielectric substance and its evaporation. In this case, a number of particles in the current cord considerably increase; newly evaporating particles are dissociated and ionized. As a result, the dense low-temperature plasma is formed in channel.

$$A_{ch} = \int_0^{V_{ch}} P_{ch} dV, \tag{4.38}$$

It acts as a working body, which is done during channel expansion against the environment. P_{ch} is the pressure in the channel and V_{ch} is the channel volume in formula (4.38).

To characterize a substance as the electric energy converter into the work, its thermodynamic properties must be known.

In [14] the equation of state of the plasma formed by some gases at temperatures $1.5 \times 10^4 < T < 2.5 \times 10^4$ K and pressures $5 \times 10^6 < P < 1.5 \times 10^9$ Pa was analyzed, and it was shown that the relationship for the thermodynamic functions can be represented in the form formally coinciding with the equation of state of the ideal gas:

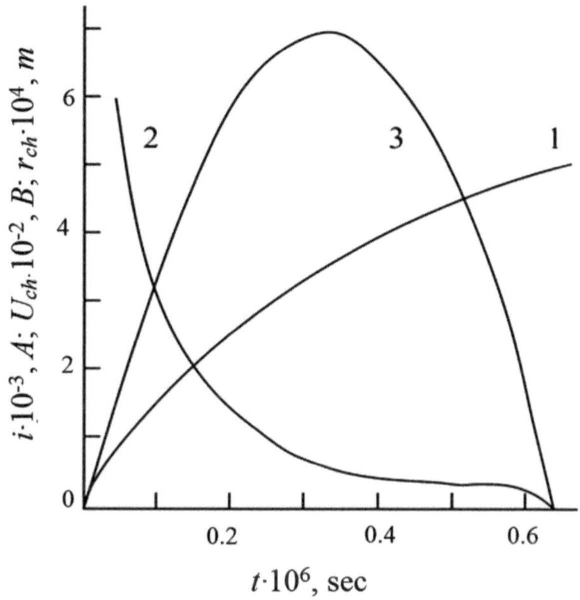

Fig. 4.14 Experimental dependences $r_{ch}(t)$ (curve 1), $U_{ch}(t)$ (curve 2), and i(t) (curve 3)

$$W_{in} = \frac{P_{ch} V_{ch}}{\gamma_e - 1}, \tag{4.39}$$

where $\gamma_e \neq sr/s_v$ (the Poisson coefficient), and an effective adiabatic index considers the dissociation and ionization processes. Such approach was used in [13] to write down the equation of state for the substance of an underwater spark.

As to the spark channel in solid dielectrics, the validity of formula (4.39) was demonstrated in [15] by calculations using experimental data. Following the scheme of calculation in [16], within the limits of the hydrodynamic model, the pressure $R_{ch}(t)$ produced in the spark channel excited in the sample made of organic glass and in KCl was calculated using finite difference methods for solving differential equations. The boundary condition was set in the form of the function of channel expansion $r_{ch}(t)$, obtained from the high-speed channel photography. The volume $V_{ch}(t)$ of the spark channel was calculated as the volume of the cylinder with radius $r_{ch}(t)$ and length equal to the interelectrode distance. The electric energy in the spark channel was determined from the waveforms of the signal current and voltage inside the gap.

The effective adiabatic exponent of the curve was determined by solving system of Eqs. (4.37–4.39). Figures 4.14 and 4.15 show the experimental dependences $U_{ch}(t)$, $i(t)$, and $r_{ch}(t)$ and the calculated characteristics $W_{ch}(t)$, $P_{ch}(t)$, and $\int_0^{V_{ch}} P_{ch} dV$.

Though the increase of γ_ε at small t corresponds to the coefficient of γ_e increase with plasma pressure [14], large errors in experimental measurements of r_{ch}, $U_{ch}(t)$, and i at $0<t<0.15$ μs do not allow us to believe that such increase has only physical nature.

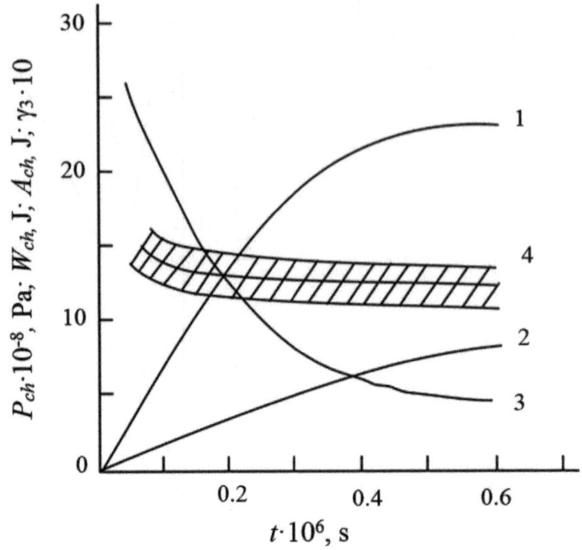

Fig. 4.15 Calculated characteristics W_c (t) (curve 1), $\int_0^{V_{ch}} P_{ch} dV$ (curve 2), P_{ch} (t) (curve 3), and $\gamma_e(t)$ (curve 4)

Small change of the parameter γ_e with time at $0.15<t<0.55$ μs, where the most part of work against the environment is done, gives us grounds to believe that the representation of internal energy W_{in} by Eq. (4.39) is sufficiently correct.

4.2 Peak Pressure in the Discharge Channel

The peak pressure in the discharge channel is proportional to $P_{ch} \equiv \frac{dN}{dt}$. This relationship follows from the energy balance equation. For linearly increasing discharge power, the constant rate of channel expansion $a_{ch} = \frac{dr_{ch}}{dt}$ and the invariable pressure P_{ch} are observed in it. Indeed, at constant pressure and without outflow from the channel mouth, energy balance Eq. (4.37), taking into account Eqs. (4.38) and (4.39), assumes the form

$$\frac{\gamma_e}{\gamma_e - 1} P_{ch} \frac{d^2 V}{dt^2} = \frac{d^2 W}{dt^2}, \tag{4.40}$$

for γ_e=const, $V_{ch} = \pi r_{ch}^2 l_{ch}$, $\frac{d^2 V}{dt^2} = \text{const}$, $\frac{d^2 W}{dt^2} = \frac{dN}{dt} = \text{const}$.

Then

$$P_{ch} = \frac{\gamma_e - 1}{2\pi l_{ch} a_{ch}^2 \gamma_e} \frac{dN}{dt}. \tag{4.41}$$

In [16], using the equation of state of water

$$P = \frac{p_0 c_0^2}{n}\left[\left(\frac{p}{p_0}\right)^n - 1\right], \quad n = 7$$

(here p_0 and c_0 are the density and sound speed in the unperturbed medium), it was demonstrated that at constant speed a_{ch} of piston expansions in water, the relationship

$$P_{ch} = \xi \rho_0 a_{ch}^2 \left(\frac{c_0}{a_{ch}}\right)^{\beta} \quad (4.42)$$

holds true.

Using the equation of state for some solid dielectrics, the pressure was calculated in the channel of impulse electric discharge and it was demonstrated that Eq. (4.43) holds true; in this case, $\xi = 1.35, 1.27, 1.20, 0.93$; $\beta = 0.29, 0.30, 0.32, 0.40$ for water, organic glass, NaCl, and granite, respectively [1].

The joint solution of Eqs. (4.41) and (4.42) allows the pressure level in the electric breakdown channel in condensed dielectrics to be estimated from the formula:

$$P_{ch} = (\xi \rho_0)^{2/(4-\beta)} \left(l_{ch} \cdot \frac{1}{2\pi} \cdot \frac{\gamma_{e-1}}{\gamma_e} \cdot \frac{dN}{dt}\right)^{(2-\beta)/(4-\beta)}. \quad (4.43)$$

Considering that the properties of concrete are close to granite which is often used as a filler in concrete, for granite it is possible to accept the following values of the coefficients: $\xi = 0.93$, $\beta = 0.40$, $p_0 = 2670\ \text{kg/m}^3$, and $c_0 = 5850$ m/s.

The effective adiabatic index of the curve is $\gamma_e = 1.20$. The calculation formula for the peak pressure of the spark discharge in concrete assumes the form

$$P_{ch} = 0.202 p_0^{0.556} c_0^{0.222} l_{ch}^{-0.444} \left(\frac{dN}{dt}\right)^{0.444}, \quad (4.44)$$

or in the dimensionless form normalized by the pressure $p_{awp} = p_0 c_0^2$ in the acoustic wave of compression

$$\frac{P_{ch}}{p_0 c_0^2} = \left(\frac{0.202}{p_0 c_0^4 l_{ch}} \cdot \frac{dN}{dt}\right)^{0.444}. \quad (4.45)$$

Using Eq. (4.32) for the segment of the curve with increasing power in the form

$$\frac{dN}{dt} = \frac{U_0^2}{2L} \cdot \frac{df_N(x)}{dx}, \quad (4.46)$$

we obtain

$$\frac{P_{ch}}{p_0 c_0^2} = \left(\frac{0.101}{p_0 c_0^4 l_{ch}} \cdot \frac{U_0^2}{L} \cdot \frac{df_N(x)}{dx}\right)^{0.444}. \quad (4.47)$$

Based on the formulas derived, we estimate the pressure in the energy deposition regime for the technology of destruction of sub-standard reinforced concrete products. In destruction of samples of concrete of brand 300 provided that the sample thickness is equal to 100 mm, the energy stored in the storage device will be 10 kJ, the voltage was 500 kV, the inductance of the discharge circuit was $L = 10$ μg, the spark constant A_{ch} was approximately 300 V s$^{0.5}$ m^{-1}, and CS_{br}=0.113.

$$f_N\max = 0.257,\ \tau_1 = 3.62,\ \frac{d\, f_N(x)}{dx} = \frac{\pi f_N^{\max}}{\tau_1} = 0.223,$$

$$\frac{dN}{dt} = \frac{U_0^2}{2L} \cdot \frac{df_N(x)}{dx} = 1.86 \times 10^{16}\ \text{W/s},$$

$$P_{ch} = 0.202\, \rho_0^{0.556} c_0^{0.222} l_{ch}^{-0.444} \left(\frac{dN}{dt}\right)^{0.444} = 5.16 \times 10^9\ \text{N/m}^2. \tag{4.48}$$

Under conditions of such intensive actions, the basic elements of the equipment—the capacitive storage devices, switches, electrodes, and electrode systems have, as a rule, limited lifetime, require considerable depreciation charges, and are unreliable in operation. In this regard, it seem urgent to investigate the effect of regimes of energy release on the efficiency of destruction of concrete and specific energy consumption in for wide range of energy deposition regimes to estimate the possibilities of application of the technology in low-intensive regimes facilitating functioning of the equipment.

References

1. Siomkin BV, Usov AF, Kuretz VI (1995) Bases of electropulse destruction of materials. Nauka, St. Petersburg, 276 p. (in Russian)
2. Dulzon A, Vajov V, Jgun D et al (2001) Electropulse wells drilling. In: International conference on pulsed power applications. Gelsenkirchen, Band 2, pp E06/1–E06/5
3. Dulzon AA, Ushakov VY (2001) State and development perspective of technological applications of high voltage high-powered pulse. In: International conference on pulsed power applications. Gelsenkirchen, March 27–29, 2001, Pap. Numb. A.02
4. Usov AF, Siomkin BV, Zinoviev NT (2000) Transient processes in installations of electropulse technology. Nauka, St. Petersburg, 160 p. (in Russian)
5. Kuretz VI, Usov AF, Zuckerman VA (2002) Electropulse disintegration of materials. Publishing House of the Kola Scientific Center of the Russian Academy of Sciences, Apatity, 324 p. (in Russian)
6. Emelin MA, Morozov VN, Novikov NP et al (1990) New methods of rock destruction. Publishing House Nedra, Moscow, 240 p. (in Russian)
7. Sulakshin SS (2004) Destruction of rocks during drilling of wells. Publishing House of Tomsk Polytechnic University, 136 p. (in Russian)
8. Vozdvizhensky BI, Rebrick BM (1989) Into the earth: exploratory drilling from the past to the future. Publishing House Nedra, Moscow, 168 p. (in Russian)

9. Toepler M (1906) Zur Kenntnis Gesetze der Gleitfunken bildung. Ann Phys (21):193–223
10. Rompe R, Weizel W (1944) Uber das Toeplersche Funken-gesetz. Z. Phys. 122(3):636–639 (in German)
11. Krivitsky EV (1972) Investigation of the resistance of the underwater spark discharge channel. Zhurnal technitcheskoy phyziki. 42(No. II):2362–2365 (in Russian)
12. Naugolnykh KA, Roy NA (1971) Electric discharges in water (Hydrodynamical description). Publishing House Nauka, Moscow, 212 p. (in Russian)
13. Okun IZ (1969) Investigation of the electrical characteristics of a pulsed discharge in a liquid. Zhurnal technitcheskoy phyziki (No. 5):837–861 (in Russian)
14. Zel'dovich YB, Riser YP (1963) Physics of shock waves and high-temperature hydrodynamic phenomena. Publishing House (Gosudarstvennoe Izdatelstvo physika—matematitcheskoy literatury), Moscow, 632 p. (in Russian)
15. Zinoviev NT, Siomkin BV (1978) To the equation of energy balance of a spark in solid dielectrics. Zhurnal technitcheskoy phyziki 48(3):624–626 (in Russian)
16. Blaznin BS, Shchegolev IA, Lozin LI et al (1983) Processing of natural stone by electric discharges. Electronnaja obrab mater (No. 1):5–7 (in Russian)

Chapter 5
Regularities of Rock Destruction in the Process of ED-Drilling of Wells

5.1 Spatiotemporal Regularities of the Breakdown in Gaps Formed by Electrodes Superimposed on One Surface of a Solid

The electric field lines in high-voltage insulating structures are parallel to the interface between the solid insulating material and the environment (atmospheric air, compressed gas, vacuum, or liquid dielectric) for the most part of the interelectrode gap, that is, the tangential component of the electric field prevails for the most part of the solid surface. In such configuration the weakest link of the electric strength is the interface along which the surface breakdown, called flashover, occurs in critical cases. This is most evident for insulating structures with water insulation (first of all, in pulsed-power systems) owing to large differences between the electrophysical properties of water and solid dielectrics [1, 2].

To ensure high reliability and high performance of high-voltage elements of the equipment, identical problems must be solved in the EDT and traditional high-voltage technology, including an increase in the flashover voltage of the insulator of actuator and prevention of its surface damage caused by high temperature in the sliding discharge channel and in some cases by the discharge channel penetration into the insulator. However, in the EDT we have to deal with the opposite problem—to provide a 100% probability and a maximal depth of discharge channel penetration into the material to be destructed/processed. Many technical operations used to prevent damage of insulation of high-voltage equipment cannot be used in the EDT based on the principle "vice versa." As already indicated above, the main method for achievement of this purpose is the use of impulses with such slope of the front that provides crossing of the voltage-time characteristics of the solid being destroyed/processed and the liquid that is in contact with it.

Whereas in the first stage of EDT development, the voltage impulse slope was recognized as the primary and practically the only factor determining the discharge trajectory in the system with superimposed electrodes, later on it became clear that

V. Y. Ushakov et al., *Electro-discharge Technology for Drilling Wells and Concrete Destruction*, https://doi.org/10.1007/978-3-030-04591-3_5

there were also other factors influencing the EDT feasibility and efficiency that depend not only on the probability of the discharge channel penetration into the solid, but also on the discharge trajectory which, in turn, determines the geometrical parameters of a stabbing crater.

The spatiotemporal regularities in the behavior of the breakdown channel were investigated for two cases typical of the ED-drillings: (a) electrodes of opposite polarity are located on the same surface of the solid being destructed, and (b) one of the electrodes is lifted above the surface. Reasons for the occurrence of such situation will be clear from a description of events that occur at the well bottom during drilling and are discussed below.

To elucidate the regularities of the breakdown process, the solid dielectric should be optically transparent and not to be destructed when the discharge channel is formed. The results shown in Fig. 5.1 were obtained for organic glass (polymethylmethacrylate). The figure illustrates the spatiotemporal breakdown schemes for both cases.

It can be seen that the schemes depend strongly on the position of electrodes relative to the solid surface (with other parameters remaining the same, including the solid, the liquid surrounding it, the voltage impulse parameters, etc.).

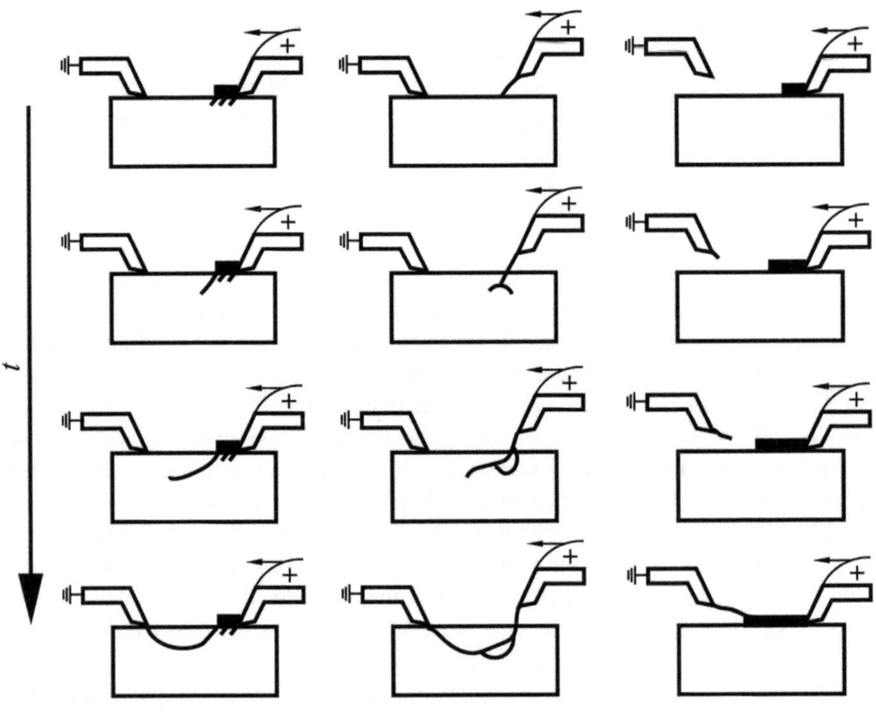

Fig. 5.1 Scetches of breakdown dynamics for positive voltage impulse polarity and indicated positions of the electrodes with respect to the solid surface

Without going into details, the breakdown can be represented as follows.

The breakdown process begins with the formation of numerous channels called by "brush discharges" near both electrodes. They evolve from the electrode in contact with the surface as sliding discharges and from the lifted electrode as a volume discharge in liquids. At a certain moment, numerous channels of incomplete breakdown develop from heads of the brush discharge channels.

In case of EDT implementation, the electrodes are short-circuited by the discharge channel developed in the solid bulk. If the solid is a heterogeneous rock, points of channel injection are mostly linked to definite mineral constituents.

The prediction of probability of channel injection and of its trajectory that determine the parameters of the stabbing crater is additionally complicated by two interesting phenomena:

1. The breakdown channel penetrates into the solid not at the *triple point*, that is, not at the point of electrode contact with the solid where the electrostatic (unperturbed) field strength is maximum, but at a distance of 2–4 mm from it; moreover, there is a certain forbidden zone of complex configuration within which the channel is not penetrated into the solid.
2. Points of breakdown channel penetration into the solid are seldom located on the shortest line connecting the electrodes; they are typically remote from it at a distance that can even exceed the interelectrode distance.

An attempt was undertaken in [3] to explain these phenomena and spatiotemporal regularities of the discharge processes in systems with superimposed electrodes based on the general concepts of breakdown physics in condensed media and of the breakdown process parameters such as discharge ignition voltage U_{ign}, rate of breakdown channel evolution in solid and liquid dielectrics, and formation of excessive volume and surface charges [4, 5]. The dependences of these parameters on the properties of solid and liquid surroundings, discharge gap geometry, and impulse parameters (shape, duration, and polarity) were considered. In this case, it should be taken into account what even in the simplest case of the symmetric system of electrodes and absolutely identical electrode shapes, the field in the discharge gap is not symmetric and the electric field gradients near the grounded electrode are lower than near the potential electrode because of the shielding action of the Earth. This additionally complicates the pattern of the breakdown processes and its interpretation.

In [3] the dependence of the drilling efficiency on the character of processes in the discharge gap in most often observed cases was explained as follows.

Scheme 5.1*a*: The discharge starts from the potential electrode at minimal U_{ign}; the "dead zone" is clearly pronounced near the potential electrode; the solid breakdown is completed before the discharge occurrence near the grounded electrode; and the volume of the stabbing crater is considerable.

Scheme 5.1*b*: The discharge starts in the liquid from the potential electrode at higher U_{ign}, than in case 5.1*a*; the discharge channel is penetrated into the solid without development of brunch discharge on the solid surface; brunch discharges have time to be formed near the grounded electrode; and the volume of the stabbing crater is maximal.

Scheme 5.1*c*: Conditions for the formation of the breakdown channel in a solid are the worst; the discharge ignition near the grounded electrode is maximally delayed from the beginning of the processes near the potential electrode; therefore, the brush discharge from the potential electrode flashes over a significant part of the interelectrode gap, and the electrodes are short circuited by the channel developing in the liquids, that is, the channel does not penetrate into the solid.

The discharge ignition is delayed for the potential electrode with negative polarity in comparison with the electrode of positive polarity; therefore, the discharge processes near the potential electrode are more intensive, and the technological effect is much weaker. In addition, the breakdown voltage is higher; therefore, the working voltage must also be higher. Thereupon, working parts with positive polarity of the potential electrode are used in the EDT and hence, it is not expedient to describe here the phenomenology of the breakdown processes for cases (schemes) with negative polarity of the potential electrode (The more so that in [6] they were described in ample detail.).

Investigations of the influence of various factors on the probability of penetration and the trajectory of the breakdown channel in a solid and hence, on the ED-drilling efficiency for model materials (organic glass, fluoroplastic, and polyethylene) are useful, but insufficient for the design of rock destructing equipment and the choice of optimal technological regimes. As demonstrated in [3], these two important parameters depend strongly on the structure (composition and crystal grain size), texture, and surface electric resistance of rocks that, naturally, cannot be reproduced when investigating model materials. In [3] it was noted that the destructibility of rocks is rather individual and cannot be predicted without experimental testing for the given level of knowledge of breakdown regularities. In this case, it must, however, be understood that in laboratory experiments it is difficult to reproduce the influence of a number of other factors manifested at the well bottom, including electrode configurations, state of the liquid (turbulence, degree of saturation with gases, elevated temperature and pressure, and contamination of the liquid by fragments of the solid being destructed), etc.

The main actuator element (drilling implement) in well drilling is the drilling head designed in the form of the system of potential and grounded electrodes uniformly distributed over the well bottom with approximately identical distances between the electrodes combined into a single whole, Fig. 5.2.

Such design allows the EDT idea to be implemented, that is, the superiority of the electrical strength of the liquid surrounding at very steep high-voltage impulse fronts to be used for destruction of solids. The most important feature of such drilling tool in comparison with traditional one used for mechanical drilling is that it is not necessity to rotate it pressing to the surface of a destructed solid, since the electric discharges are automatically distributed over the gaps. When one of the gaps (for example, number 1 in Fig. 5.3) breaks down, the stabbing crater formed in the rock is filled with flushing-out liquid, thereby leading to its electric strengthening.

The subsequent discharges occur alternately in other gaps until the destruction of rock in all gaps at the bottom of the well. After that the drilling implement moves to

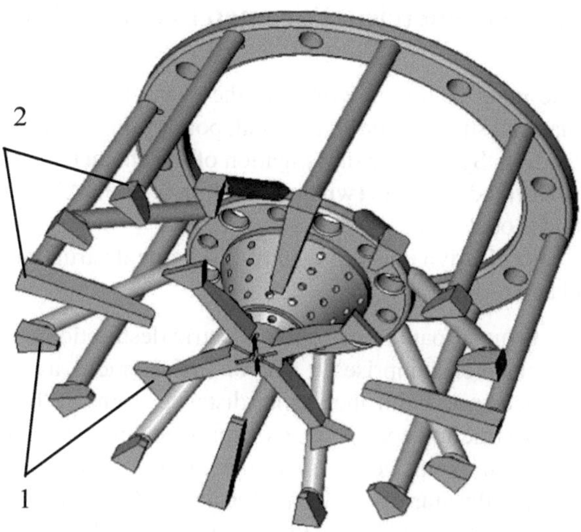

Fig. 5.2 Sketch of the drilling head for drilling wells of large diameter comprising high-voltage electrodes 1 and grounded electrodes 2 (The design was suggested in [3, 7])

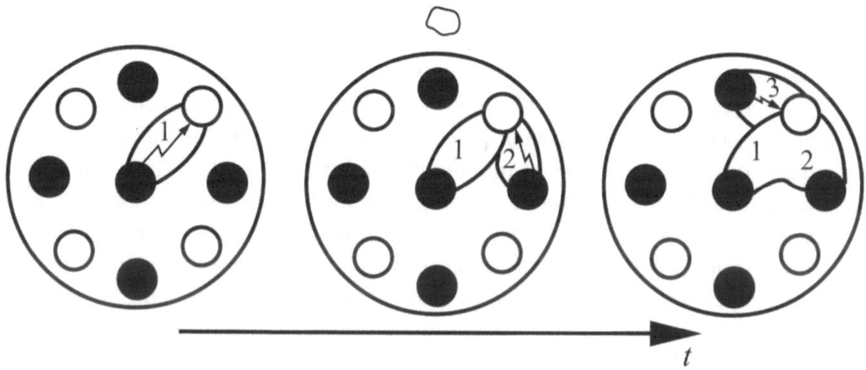

Fig. 5.3 Sequence of breakdowns of interelectrode gaps of the drilling head (Filled circles are high-voltage electrodes)

a new position in which the breakdown voltage for all interelectrode gaps becomes approximately identical. This cycle is repeated with periodicity preset by the parameters of technological regime of the drilling implement.

Almost any shape of the well bottom–round, rectangular, etc.—can be obtained by changing the geometrical parameters of the drilling head (the shape and the number of electrodes, the distance between them, and their mutual arrangement). The bottom area can also be changed in wide limits (According to [3], up to more than ten square meters).

Arrangement of the electrodes along the perimeter of the well bottom allows prospecting wells with core selection to be drilled.

5.2 Geometrical Parameters of a Slabbing Crater

The main factors determining the character and the dynamics of cracking are the material structure (single crystal, polycrystalline, or amorphous) and the intensity of loading. Systematic investigation of the character and dynamics of destruction were performed for glass (window and mirror one), AHC, silicon single crystals, PMMA, and model rocks.

For the dynamics of electro-discharge destruction, the following regularities are characteristic:

1. General pattern of solid dielectric destruction under the action of the discharge channel comprises a star of radial cracks as a main element. Their number decreases with increasing distance from the breakdown channel. The volume destruction zone is poorly expressed, and ring cracks, observed during explosion of an explosive, are as a rule absent in this case. The volume destruction zone and the star of cracks are formed under the action of wave perturbations. In the final stage, including the phase of the final stop (equilibrium) of radial cracks, the determining mechanism of energy transfer in the crack mouth is the force impact of the discharge channel (zone of plastic deformations); the energy necessary for crack growth is supplied to the mouth by the Rayleigh waves.
2. The characteristic feature of brittle material destruction is almost constant rate of crack growth, up to the moment of its stop. In silicate glass it reaches 1500 m/s, which corresponds to 0.38 of the longitudinal wave velocity. This regularity is characteristic for a wide range of energy regimes. In the AHC, separate cracks are registered only at low energy content in the channel, and the rate of their growth is quasi-constant, of the order of 0.38 of the wave velocity. In rocks with high dislocation concentrations, destruction is spread by merging of separate microcracks, and its rate corresponds to the velocity of elastic wave propagation. To investigate the destruction process, indirect methods are employed using

   - sealed electrodes when the discharge channel in the sample is formed at a fixed distance from the surface,
   - optical high-speed photo registration to determine the time during which electrical breakdown products emerge onto the surface,
   - oscilloscopic recording of the dynamics of variation of the electric discharge channel resistance under assumption that its sharp decay due to unloading corresponds to the moment of crack emergence onto the surface.

   The results obtained for a number of rocks confirmed the mechanism of destruction spreading with the acoustic wave front.
3. The effect of ED destruction of materials with the same energy consumption depends on the character of energy liberation in the discharge channel. The destruction efficiency can be judged from such parameters, as the maximal crack length, total crack length and area, area of the cracking zone, etc. The most typical dependence of the above-indicated parameters on the rate of energy liberation for its constant value is a curve with a maximum. Depending on the character

of the material (brittle or plastic), the maximum is significantly shifted toward lower or higher powers.

The effect of unloading of the discharge channel (channel energy liberation through its mouth and cracks emerging onto the surface) leads to a reduced effective time of its loading; that is why the discharge gap length and the discharge penetration depth affect significantly the character of the dependence of the destruction on the discharge power. For long gaps the dependence peculiar to plastic materials is observed for rocks, and for short gaps, the dependence peculiar to brittle materials is observed.

Optimization of the regime of energy input into the discharge channel provides considerable prospects for improvement of energy characteristics of the ED-process. The physical essence of the method makes it possible to ensure lower energy consumption for destruction in comparison with the traditional methods.

The energy consumption for the destruction of some Kola Peninsula rocks from the data obtained for a single breakdown of the indicated samples is given in Table 5.1 [8]. The characteristics of drilling of more than 40 rocks with three-electrode drilling head 50 mm in diameter were presented in [3]. The drill ability for rocks, in mm/J, was the following: 1.8 for frozen clay, 1.1 for gravel, 1.05 for sandstone, 0.97 for dolomite limestone, 0.87 for shale, 0.86 for marble, 0.76 for quartzite, 0.6 for coarse-grained granite, 0.58 for granite-porphyry, 0.48 for microdiorite, 0.3 for gabbro, 0.1 for tuff slate, and 0.04 for galena-sphalerite ore (Table 1.1) (Magnetite was not available for ED-drilling.).

By energy efficiency, the EDT is noticeably preferable over the majority of other methods of rock destruction from one free surface (including drilling, cutting, and removal of the surface layer). Moreover, the EDT has considerable, yet unrealized, potential for improving the energy efficiency. Among the methods of its implementation, increase in discharge gap lengths and application of special schemes of impulse shaping, capable of more than doubling the efficiency of VIG energy transfer to the discharge channel, should be mentioned. For example, Table 5.2 presents the urtite destruction characteristics as functions of the increasing discharge gap. The destruction productivity increases by an order of magnitude, and the energy consumption decreases by more than a factor of 2.5 when the gap length increases from 15 to 55 mm.

The important indicator of the electric spark efficiency as a working tool of the EDT is the size (first of all, depth) of the stabbing crater formed by each discharge. The shapes and volumes of the stabbing craters are extremely diverse. The volume of the stabbing craters undergoes smaller fluctuations. The length, width, and depth of the stabbing crater characterize its maximal sizes in the given series of experiments.

The empirical relationship between the stabbing crater depth H and the distance between the electrodes S is described by the expression $H = 0.32S$ valid under certain restrictions [3, 9]. In the destruction of sandstone and granite, the relationship between H and S for interelectrode distances up to 70 mm can be described by the expression

Table 5.1 Power consumption for the destruction of Kola Peninsula rocks

Rock	Hardness, f_P	Energy consumption, J/cm^3 (kg/cm^3)
Microcline pegmatite	12–14	185 (19)
Amphibole-biotite gneiss	14–16	195 (19)
Sericite-quartz shale in porphyrites	4–8	285 (29)
Apatite ore	6–8	295 (30)
Talc-carbonate rock	6–8	320 (32.8)
Urtite	10–12	365 (37.5)
Foyaite	11–15	390 (40)
Hornblende gabbro	16–18	410 (42)
Micro quartzite	Up to 20	520 (53)
Ore-free olivine pyroxenite	14–6	550 (56.5)
Porphyrite	12–14	640 (65)
Sericite-quartz-chlorite rock	12–14	640 (65)
Vein plagioclase		765 (78)
Sericite-quartz schist over sedimentary rocks	6–8	805 (82)
Talc rock	2	805 (82)
Serpentinite	8	1040 (106)
Diabase	18–20	1070 (109)
Sericite-quartz rock		1335 (36)
Garnet-biotite gneiss	10–14	1725 (175)
Serpentinized verlite		Not destroyed
Pyroxenite olivine with mineralization		–
Solid polymetallic ore		–

Table 5.2 Energy characteristics of ED-destruction of urtite for the indicated discharge gap lengths

S, mm	W, J	V, cm^3	W_{spec}, J/cm^3
15	410	0.54	760
30	1200	2.8	425
40	1200	3.2	375
55	1470	5.3	280

$$H = H_0 + kS, \quad (5.1)$$

where H_0 is the penetration depth for $S = 20$ mm and k is the experimental coefficient. For granite, $k = 0.17$ and $H_0 = 6.34$.

Experiments demonstrated that the discharge channel penetration depth is stochastic in nature. Significant deviations of the results obtained by different authors are caused by different employed experimental procedures. The random character of the discharge channel trajectory in the rock leads to the variance of the maximal penetration depths. In rocks it is difficult to determine the discharge channel penetration depth. It is usually estimated from the depth of the stabbing crater which is greater than the channel penetration depth. It is well known that with increasing distance between the electrodes, the absolute depth of the discharge channel penetration into the solid (H) increases, but its relative depth ($h = H/S$) decreases [3].

The range of variations of the channel penetration depths and destruction parameters of different rocks is sufficiently wide and considerably spread, which is caused by the composition and structure of the rock. To reveal the general regularities characterizing the depth of the discharge channel penetration into the rocks, the geometrical sizes of stabbing craters were investigated for the ED-destruction of granite samples in water and transformer oil with distances between the electrodes in the range 10–300 mm, energy input in the range 10–110 J/mm, and the number of acting impulses from 2 to 15 [3, 10].

The knowledge of the dynamics of stabbing crater sizes allows regimes of ED-drilling, cutting, and surface layer removal from the stone and concrete to be optimized.

Figure 5.4 shows the dependences of the relative penetration depth h into granite on the distance S between the electrodes. (Here $h = H/S$.) Values of H were measured under the action of three voltage impulses.

Even in this case, considerable spread of the destruction depths was observed, especially at small S values. For example, for $S = 10$ mm, the minimum and maximum H values might differ twice. A 30-fold increase in the distance between the electrodes (from 10 to 300 mm) led to the decrease of the relative destruction depth by a factor of 2.33.

The dependences of the maximal relative destruction depths averaged over measurement series on the distance between the electrodes can be described by the expression

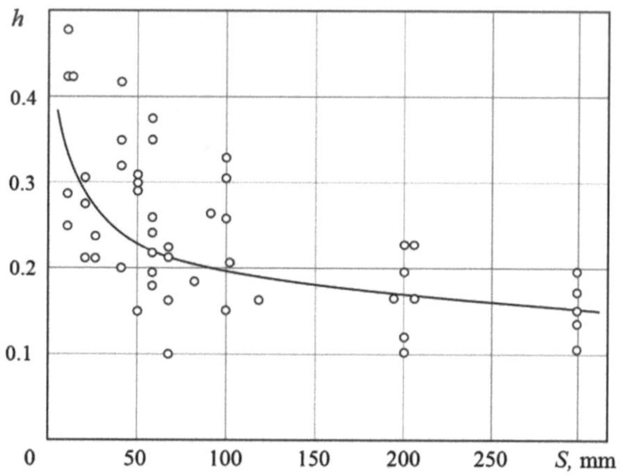

Fig. 5.4 Dependence of the relative destruction depth on the distance between the electrodes for $n = 3$ pulses (The solid curve shows results of calculation from Eq. (5.2), and the open circles show experimental data)

$$h = h_{10} S^{-a}, \tag{5.2}$$

where $h_{10} = 0.36$ is the average maximal relative destruction depth for $S = 10$ mm, and $a = 0.25$ is the empirical coefficient.

Figure 5.5 shows the dependences of the relative destruction depth on the number of impulses acting on the rock for the indicated distances between the electrodes.

At large S (exceeding 100 mm), the samples were destructed under the action of 6–7 impulses, and their tests were stopped. The general regularity was the saturation of h with increasing n. Thus, h increased to 0.6 for $S = 10$ mm, that is, by a factor of 1.69 when n increased from 3 to 15 impulses. In particular, this is one of the reasons of different h values reported in publications of different authors.

The main reasons for increase in $h = f(n)$ are cracks and incomplete breakdown channels remaining from previous impulse actions.

The difference between the destruction depth and the depth of discharge channel penetration into the rock is illustrated by Fig. 5.6, which shows the dependences of h on the energy input W_s per unit length of the interelectrode distance for the indicated S values under the action of three voltage impulses. For all S values, the destruction depth increased with energy input the stronger, the smaller is the distance between the electrodes. Since the energy input is determined by the energy stored in the VIG, to increase the rate of rock destruction, it is necessary to increase W_s. This is one more reason for different H values reported by different authors.

These parameters for the indicated interelectrode distances, two values of the energy input $W_s = 20$ J/mm and $W_s = 49$–60 J/mm, and three acting impulses are presented in Table 5.3.

Based on the results presented in Table 5.3, the following tendencies can be revealed in the change of the length and width of the stabbing crater referred to the interelectrode distance:

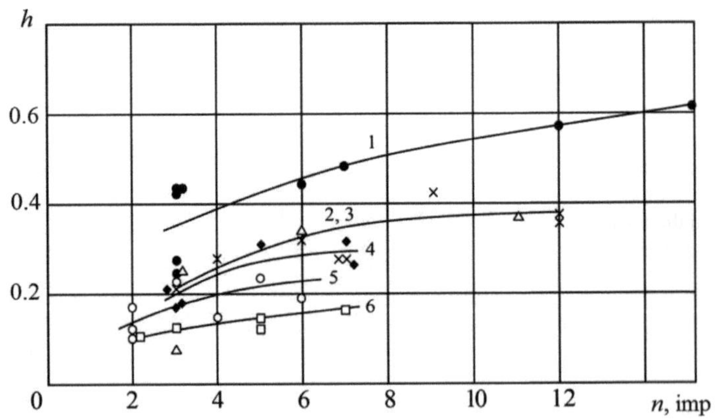

Fig. 5.5 Dependence of the relative destruction depth on the number of acting impulses for the interelectrode distance $S = 10$ (1), 20 (2), 50 (3), 100 (4), 200 (5), and 300 mm (6)

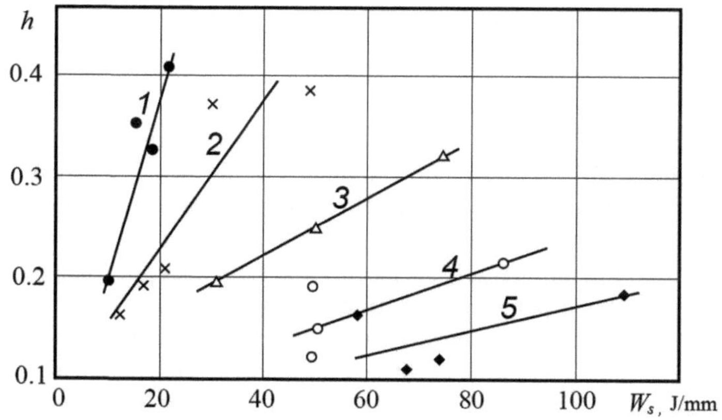

Fig. 5.6 Dependence of the relative destruction depth on the energy input for the interelectrode distance $S = 40$ (*1*), 60 (*2*), 100 (*3*), 200 (*4*), and 300 mm (*5*) and $n = 3$ impulses

Table 5.3 Relative maximal sizes of stabbing craters in granite for $n = 3$ impulses

W_s, J/mm	S, mm	l/S	B/S	H/S
23.1	40	1.62	0.92	0.34
20.4	60	1.41	1.21	0.30
20.0	90	1.13	1.00	0.20
20.1	120	0.78	0.62	0.16
52.0	48	1.56	1.39	0.31
50.0	60	1.50	1.15	0.38
60.0	100	1.83	1.21	0.38
52.5	120	1.20	0.63	0.21
49.0	140	1.00	0.63	0.17

(1) The relative sizes of the stabbing craters decreased with increasing distance between the electrodes given that the energy input remained unchanged,
(2) The increase of the energy input led to an increase in the relative sizes of the stabbing craters,
(3) Absolute values of the length and width of the stabbing craters might exceed the distance between the electrodes.

The character of changes of the stabbing crater depth revealed for a large volume of statistical data is shown in Fig. 5.4.

The increase of the distance between the electrodes was accompanied by the disproportionate increase in the breakdown voltage (see Eq. (2.9) and Fig. 2.7). Hence, the average electric field strength ($E_{av} = U_{br}/S$) decreased with increasing distance at the required working voltage; as a consequence, the maximal length of the breakdown channel and the destruction volume decreased.

5.3 Influence of the Energy Input Dynamics on the Destruction Characteristics

The productivity of ED-destruction is determined by the average destruction volume per single impulse V_1. The destruction volume depends on the variety of factors the main of which are the distance between the electrodes S, discharge penetration depth H, probability of discharge penetration into the solid P, the acting impulse energy W, and time of its liberation t_{lib}. If the electrode system does not change its position, the destruction volume per single impulse, having reached a maximum, starts to decrease, since the stabbing crater depth reaches a maximum, and the destruction represents expansion of its side walls. The destruction process practically stops when the stabbing crater reaches its maximal sizes. Hence, the number of impulses applied to the electrode system is the important factor determining the efficiency of ED-destruction.

Detailed investigations of the regularities in the behavior of the main characteristics of ED-destruction, including productivity of a single discharge, were performed for granite samples most often encountered in drilling of very strong rocks.

The breakdown voltage U_{br}, the time to breakdown t_{br}, and the parameters of stabbing craters—the maximal depth H, width B, length l, and destruction volume V_1—were measured during testing. The current in the discharge circuit was always oscillatory in character. It was estimated from the amplitude and period of oscillations. The period of current oscillations t_{osc} for different generators changed in the range 5.6–8.0 μs.

The destruction was characterized by the following parameters:

- Relative destruction depth of stabbing craters $h = H/S$,
- Energy stored in the VIG,
- Destruction productivity per single impulse $Q = V_\Sigma/n$, where n is the number of applied impulses,
- Specific energy consumed for rock destruction and formation of stabbing craters $W_{spec} = W_\Sigma/V$, where $W_\Sigma = nW$,
- Energy input per unit length of the interelectrode gap $W_s = W/S$.

Under conditions characteristic for ED-drilling, an increase in the number of impulses acting on the rock led to a change in the destruction productivity, Fig. 5.4. The maximal productivity of single impulse was observed under the action of from 3 to 5 impulses. In this case, the maximal productivity increased by a factor of 32.7 when S increased from 50 to 300 mm. Hence, to increase the destruction productivity, it is necessary to increase the distance between the electrodes. The data on the maximal productivity shown in Fig. 5.7 cannot be considered optimal since the energy stored by the generator was chosen arbitrarily for each S.

Figure 5.8 shows the dependences of the productivity (curve *1*) and specific energy consumption (curve *2*) on the energy stored in the VIG for $S = 60$ mm and $n = 3$ that corresponded to the range of variation of the energy input W_s.

Fig. 5.7 Dependence of the destruction productivity on the number of impulses for the interelectrode distance $S = 50$ (*1*), 100 (*2*), 200 (*3*), and 300 mm (*4*)

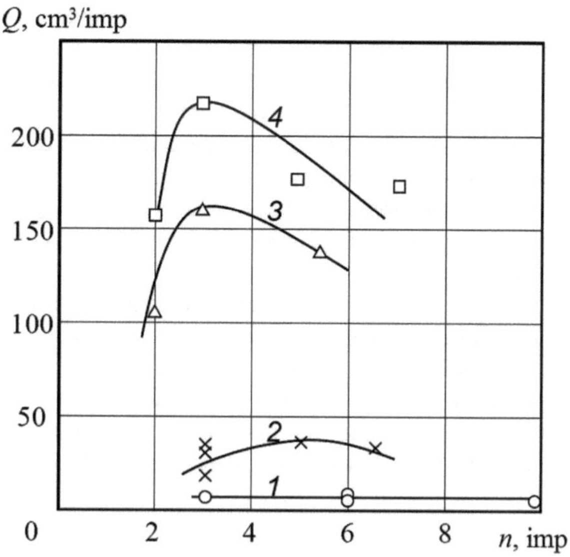

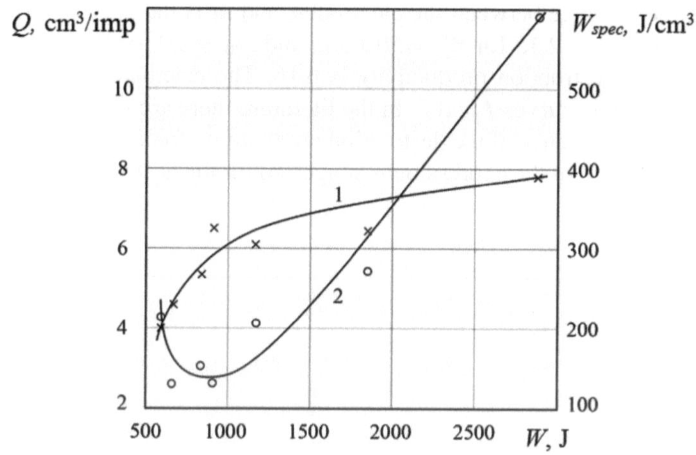

Fig. 5.8 Dependence of the destruction productivity and specific energy consumption on the energy stored in the VIG: (1) $Q = f(W)$ and (2) $W_{spec} = f(W)$ (curves here show the results of calculation from empirical Eqs. (5.3) and (5.6), and symbols are for the experimental data)

An increase in W causes a continuous increase in the productivity, especially fast for small W. In this case, the specific energy consumption has a minimum at $W_{spec} \cong$ 650–900 J. If we optimize the destruction process for the specific energy consumption, the given W value will be optimal, and the optimal energy input for $S = 60$ mm will be $W_s = 10.8$–15.0 J/mm.

It can be assumed that the energy input for $S = 60$ mm is not a constant for other S values, because the absolute penetration depth increases with S, and larger impulse energy is required for the formation of stabbing craters, given that the duration of current flow remains unchanged.

From the viewpoint of optimization of W and W_s in the entire range of S of interest for the EDT, it is expedient to compare the productivity and the specific energy consumption given that the energy input remains unchanged. Investigations were performed for energy input changing from 3.6 to 100 J/mm. According to [11, 12], for a comparison we have chosen the most reliable values of the energy input in the entire range of variation of distances between the electrodes equal to 20 and 50 J/mm.

Symbols in Fig. 5.9 show experimental values of the destruction productivity as functions of the distance between the electrodes for the energy input equal to 50 (curve *1*) and 20 J/mm (curve *2*) and three acting impulses.

The destruction productivity increases by several hundred times when the distance between the electrodes increases from 10 to 200 mm for these values of the energy input. This dependence can be described by the expression

$$Q = 0.16S^{ai}, \tag{5.3}$$

where S is the distance between the electrodes, and ai is the exponent determined experimentally: $a_{50} = 2.15$ for $W_s = 50$ J/mm and $a_{20} = 2.0$ for $W_s = 20$ J/mm; for $S = 10$ mm, the destruction productivity is 0.16. The calculated dependences are shown in Fig. 5.6 by curves *1* and *2*. In the literature there are no empirical relation describing the dependence of the destruction productivity for a single impulse on the main acting factors, but the expression is suggested for the destruction volume of the form:

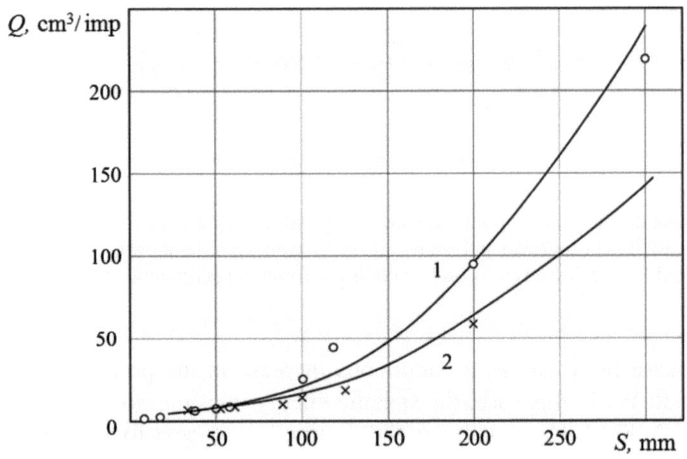

Fig. 5.9 Dependence of the destruction productivity on the distance between the electrodes for the same energy input Here symbols show experimental data, and solid curves show results of calculation) for $W_s = 50$ (curve 1) and 20 J/mm (curve 2)

$$V_1 = v_0 S^B, \tag{5.4}$$

where v_0 and B are the coefficients that differ for different rocks [13]. In particular, for urtite

$$V_1 = 0.27 S^{1.5}. \tag{5.5}$$

The equation is valid for S changing from 15 to 55 mm and arbitrary values of W and n.

From Fig. 5.9 and Eqs. (5.3) and (5.4) it follows that a very efficient method of increasing the destruction productivity is the increase of the distance between the electrodes. However, the value of the energy input must be optimized for each interelectrode distance, which can be seen from curves *1* and *2* in Fig. 5.9. For $S > 60$ mm, the decrease in the energy input from 50 to 20 J/mm leads to the corresponding decrease of the destruction productivity by a factor of 1.58 for $S = 200$ mm. It can be assumed that the energy input $W_s = 50$ J/mm is excessive for $S < 60$ mm, since the destruction productivity practically did not decreased even for $W_s = 20$ J/mm.

One of the main characteristics of rock destruction is the specific energy consumption W_{spec}. It is well known that the energy consumption decreases with increasing distance between the electrodes. This dependence is proportional to $S^{(1-b)}$. According to [3, 9], $b = 1.6$ and $W_{spec} \equiv S^{-0.6}$ for urtite. In these works, it was pointed out that for correct evaluation of the exponent, the process of energy liberation in the discharge channel should be optimized. The dependence $W_{spec} = f(S)$ is shown in Fig. 5.10 for sufficiently wide range of variations of S at constant energy inputs $W_s = 50$ J/mm (curve *1*) and $W_s = 20$ J/mm (curve *2*). A sharp decrease in the energy consumption is observed with increasing S, especially at small distances from 10 to 50 mm (curve *1*).

Curve 1 can be described by the dependence of the form

$$W_{spec} = W_{10}/S, \tag{5.6}$$

where W_{10} is the specific energy consumption for $S = 10$ mm. For example, for $W_s = 50$ J/mm, $W_{10} = 3000$ J cm/cm^3. For other W_s values, W must be determined experimentally.

An analysis of the dependences shown in Figs. 5.9 and 5.10 and described by Eqs. (5.3) and (5.6) demonstrates that for each interelectrode distance, there exists an optimal value of the energy input and of the energy stored by the VIG at which the destruction of rock is most efficient.

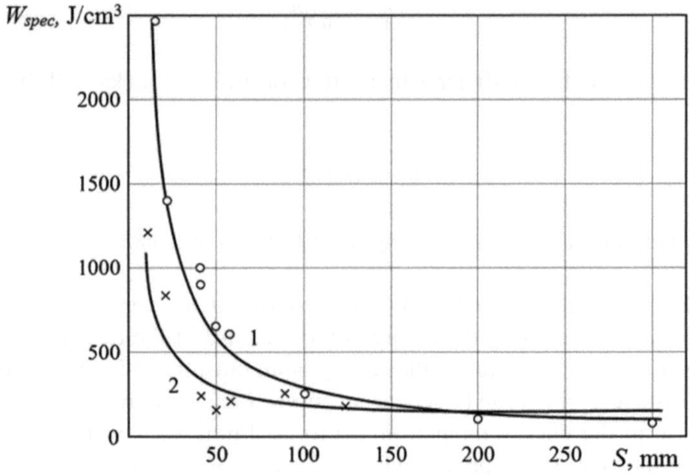

Fig. 5.10 Dependence of the specific energy consumption on the distance between the electrodes for $W_s = 50$ (curve 1) and 20 J/mm (curve 2) (Here symbols are for the experimental data, and solid curves are for the results of calculation)

5.4 Combined Effect of Elevated Pressures and Temperatures on the ED-Destruction of Rocks

The search for fields of the most efficient application of EDT drilling of wells stimulated continuation of research of ED-destruction of rocks under the action of elevated pressures and the temperatures performed for the first time in Tomsk Polytechnic University still in the 60s of the XXth century. Using samples of three rocks—quartzite, felsite-porphyry, and marble as well as organic glass in transformer oil at pressures of 2.5–15 MPa for a distance between the electrodes of 11 mm, it was established that with increasing pressure, the size of the cracking zone in organic glass decreases, that is, the number of radial and ring cracks and their lengths decreased. The depth of discharge channel penetration into solid dielectrics and rocks was independent of the external hydrostatic pressure up to 15 MPa and equal to (0.27–0.50)*S*. The decrease of the destruction productivity and the increase of the energy consumption by a factor of 1.3–1.4 were observed at pressures in the range 2.5–15 MPa.

Later, the breakdown and destruction of rocks under the joint effect of the elevated pressures and temperatures was investigated in the Versa Pro drilling solution. In addition, control experiments on the breakdown and destruction of the same rocks under the action of only the elevated pressure were performed.

The action of the elevated temperatures and pressures on the flushing-out liquid and rock, changing their electric strength, can also change the discharge channel trajectory and hence, the stabbing crater volume.

Investigations were performed in the chamber the pressure in which could be increased up to 350 atm and the temperature could be increased up to 120 °C under application of impulse voltage up to 400 kV.

The rocks that differed by their physicomechanical characteristics, composition, structure, and conditions of their formation were investigated, including sandstone, limestone, and granite. Samples with sizes of 120 × 80 × 40 mm were cut from a massive piece of rock. In the chamber the samples were fixed on a dielectric substrate, and the rod electrodes, one of which was grounded, were superimposed on a larger surface. The distance between the electrodes was 20 mm. The chamber was filled with Versa Pro drilling solution.

The VIG allowed the energy up to 500 J to be stored and the voltage impulses up to 400 kV to be generated. The breakdown was initiated on the front of the voltage impulse of positive polarity with duration of 0.22 μs.

The length *L*, width *B*, depth *H* and volume *V* of the stabbing crater were measured under application of five high-voltage impulses. This was caused first, by the probabilistic character of the discharge channel penetration into the solid (for a 20-mm distance between the electrodes, the probability of channel penetration into granite is 75–85%); second, by the fact that the highest destruction efficiency was observed under application of three-six impulses. These results testify to the strong effect on the destruction of the pressure and the weak effect of the rock type in this range of pressures. However, if we compare the destruction characteristics at pressures from the atmospheric one to 35.0 MPa, the effect of the material becomes obvious and first of all, of the mechanical strength, Table 5.4. The destruction depth was independent of the pressure and rock type when the discharge channel penetrated into the rock.

Figure 5.11 shows the dependences of the destruction depth *H* under the simultaneous action of the pressure and temperature (in this figure and below, the figures adjacent to the symbols indicate the temperature in degrees Celsius). Despite significant variance of the results obtained, it is possible to assert that the destruction depth is independent of *P* and *T* in the chosen range of their variations. In this case, the average values were close for the examined rocks and equal to $\bar{H} \approx (0.27 - 0.32)S$ in the entire range of variations of the pressure and temperature, that is, were approximately the same as under normal conditions on the earth surface. The crater depth depended weakly on the electrophysical and elastoplastic properties of rocks. At present there are no convincing proofs of the influence of electrophysical properties of rocks and liquids on the depth of the discharge channel penetration.

Table 5.4 Relative characteristics of the ED-destruction productivity at elevated pressures and temperatures

Rock	Pressure		Pressure + temperature	
	$Q_{0.1}/Q_{35}$	$Q_{2.5}/Q_{35}$	$Q_{0.1}/Q_{35}$	$Q_{2.5}/Q_{35}$
Granite	9.2	1.65	10.9	4.0
Limestone	7.4	1.65	5.3	2.1
Sandstone	3.5	1.50	4.1	2.1

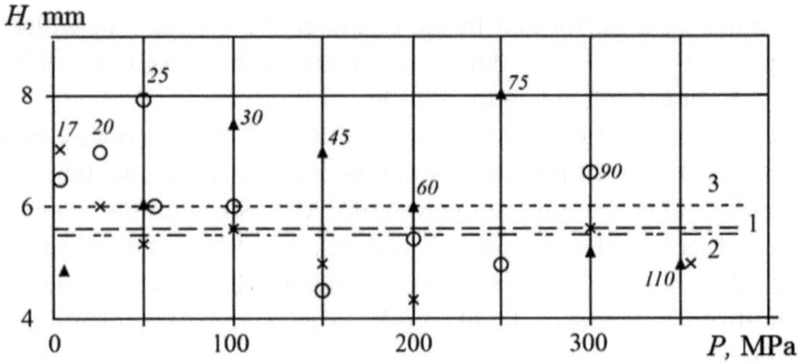

Fig. 5.11 Dependences of the destruction depth on the pressure and temperature. Here curve 1 is for sandstone (empty circle), curve 2 is for limestone (times), and curve 3 is for granite (filled triangle); figures adjacent to the symbols indicate the temperature

Whereas the stabbing crater depth changed weakly with increasing pressure and temperature, its width and length considerably decreased, thereby leading to a decrease in the volume V and the productivity Q.

The productivity is determined most correctly from the volume of the stabbing crater since it is measured with accuracy no less than 5% and possesses a smaller variance. Figure 5.12 illustrates the dependence of the average destruction productivity recalculated per single impulse for simultaneous increase in the pressure and temperature. The destruction volume fast decreased for relatively low pressure (2.5–5.0 MPa) and temperature close to room one. Hence, the decisive role in Q decrease is played by the pressure. In this case, the physicomechanical properties of rocks were more important.

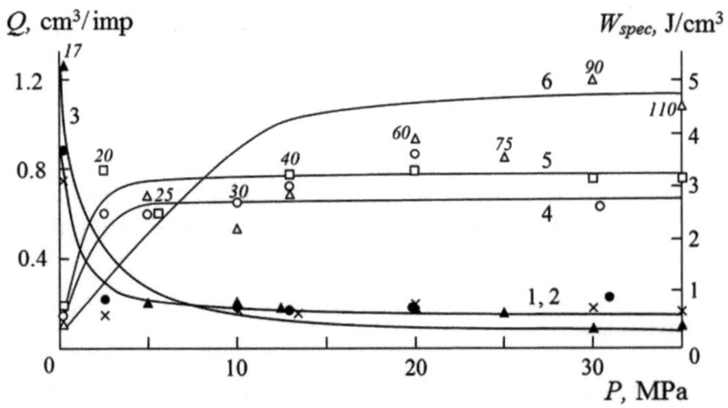

Fig. 5.12 Dependences of the destruction productivity (curves 1–3) and energy consumption (curves 4–6) on the pressure and temperature for sandstone (curves 1 and 4), limestone (curves 2 and 5), and granite (curves 3 and 6)

Table 5.5 gives values of the relative stabbing crater volume for three examined rocks. Under normal atmospheric conditions ($P = 0.1$ MPa and $T = 17$ °C), the destruction volumes for sandstone and limestone were significantly less than for granite. However, at the pressure $P = 5$ MPa, the relative decrease in Q for granite was much greater than for sandstone and limestone. This peculiarity was retained with further increase in P up to 35 MPa (Fig. 5.12; Table 5.6). It should be noted that absolute values of the destruction volumes for different rocks at high pressures and temperatures came closer together, and at $P \sim 10$ MPa became almost identical.

The energy consumed for the destruction at elevated pressures and temperatures fast increased at pressures up to 10 MPa and temperatures up to 30 °C (Fig. 5.12; Table 5.7). Further increase in the pressure and temperature affected weakly the energy consumption and hence, the destruction productivity. The greatest relative increase in the power consumption was characteristic for granite.

Simultaneous increase in the pressure up to 35 MPa and temperature up to 110 °C led to a 5–11 fold decrease of the destruction productivity and the increase of the energy consumption for the examined rocks. Their most considerable changes occurred at pressures in the range 0.1–5 MPa and temperatures up to 30 °C. This is primarily caused by the increase of the pressure. Further increase in the pressure and temperature affects weakly the productivity and the energy consumption.

The significant decrease of the ED-destruction productivity and increase of the energy consumption calls into question the effectiveness of EDT application for

Table 5.5 Relative change of the destruction volume

Rock	Relative change of Q		
	$P = 0.1$ MPa $T = 17$ °C	$P = 5$ MPa $T = 25$ °C	$P = 35$ MPa $T = 110$ °C
Granite	1.0	6.5	11.2
Limestone	0.73	3.7	4.6
Sandstone	0.62	4.5	5.6

Table 5.6 Relative change of the energy consumption

Rock	Relative change of W_{spec}		
	$P = 0.1$ MPa $T = 17$ °C	$P = 5$ MPa $T = 25$ °C	$P = 35$ MPa $T = 110$ °C
Granite	1.0	6.6	11.4
Limestone	1.65	4.6	4.6
Sandstone	1.36	4.5	5.4

Table 5.7 Characteristics of granite destruction

S, mm	P, MPa	T, °C	Q, cm^3/impulse	W_{spec}, J/cm^3
10	25	85	0.02	52,500
20	25	85	0.11	9545
30	25	84	0.31	3500

drilling of deep wells. However, analogous effect takes place for the traditional methods of drilling of deep wells. For example, the rate of diamond or thermo-shell drilling of rocks with VI–IX hardness according to the Mohs scale decreased by a factor of 12–16 when the well depth increased to 2500 m. As demonstrated above, the ED-destruction productivity of rocks significantly increased with distance between the electrodes. To elucidate whether this effect is retained at simultaneous increase of high pressures and temperatures, the evaluation experiment was carried out. The distances S changed in the range 10–30 mm. Table 5.7 presents the results averaged over 3–5 experiments for each distance.

From Table 5.7 it follows that with increasing distance S, the destruction productivity fast increases, and the energy consumption decreases. In the above-indicated range of distances, the productivity can be described by expression

$$Q = Q_0 S^{2.5}, \tag{5.7}$$

where $Q_0 = 0.02$ cm^3/impulse is the productivity for $S = 1$ cm, that is, Q increases with S greater than it follows from Eq. (5.3). Since deep drilling is performed for wells with large diameters, the distance between the electrodes in deep drilling can be made greater than 30 mm. Considerable advantages of the ED-drilling of deep wells over the traditional mechanical methods were confirmed by tests performed by the customer (in Norway) under field conditions.

The results of laboratory investigations and field tests generalized in [6] and [10] and in this monograph have allowed us to indicate the following main EDT advantages:

(1) Relatively high destruction efficiency with increased distances between the electrodes,
(2) Significant well penetration without replacement of the drilling head,
(3) Possibility of application of a submerged high-voltage impulse generator due to large diameter of wells.

The investigations described in Sect. 5.4 were supported in part by Schlumberger Ltd. and the U.S. Civilian Research and Development Foundation (CRDF).

References

1. Ushakov VY (2004) Insulation of high-voltage equipment. Springer, Berlin, Heidelberg, 421 p
2. Ushakov VY, Klimkin VF, Korobeynikov SM (2007) Impulse breakdown of liquids. In: Ushakov VY (ed). Springer, Berlin, Heidelberg, 397 p
3. Siomkin BV, Usov AF, Kuretz VI (1995) Bases of electropulse destruction of materials. Nauka, St. Petersburg, 276 p (in Russian)
4. Vershinin YN (2000) Electron-thermal and detonation processes in the electric breakdown of solid dielectrics. Publishing house of Uro RAN, Ekaterinburg, 259 p (in Russian)

5. Gournay P, Lesaint O (1994) Evidence of the gaseous nature of positive filamentary streamers in liquids. In: Conference on electrical insulation and dielectric phenomena. Ann report № 4, pp 834–839
6. Ushakov VY, Dulzon AA (2001) Performance capability of technological installations using the electrical discharge or conductor electrical explosion energy. In: International conference on pulsed power applications, Gelsenkirchen, 27–29 March, p A.03
7. Usov AF, Siomkin BV, Zinoviev NT (2000) Transient processes in installations of electropulse technology. Nauka, St. Petersburg, 160 p (in Russian)
8. Pattini G Simoni L (1971) Durata di vita dei dielettrici sottaposti a tensione alternata. L'elettratecnica 58(1):25–38
9. Kuretz VI, Usov AF, Zuckerman VA (2002) Electropulse disintegration of materials. Publishing House of the Kola Scientific Center of the Russian Academy of Sciences, Apatity, 324 p (in Russian)
10. Vajov VF, Zinoviev NT, Ushakov VY (2016) Electrodischarge technology for drilling wells and fracture of reinforced concrete products: monograph. In: Ushakov VY (ed). Publishing house of Tomsk Polytechnic University, 312 p (in Russian)
11. Boev S, Vajov V, Jgun D et al (1999) Destruction of granite and concrete in water with pulse electric discharges. In: 12 IEEE international pulsed power conference, USA, Monterey, vol 2, pp 1369–1371
12. Fabiani D, Simoni L (2005) Discussion on application of the Weibull distribution to electrical breakdown of insulating materials. IEEE Trans Dielectr Electr Insul 12(1):11–16
13. Boev S, Vajov V, Jgun D et al (2000) Research of conditions of material destruction by electric pulse discharge. In: 4 Korea–Russia international symposium of science and technology. Part 3. Korus 2000, Korea, Ulsan, pp 346–349

Chapter 6
Insulating Design in EDT Installations

6.1 Requirements for Insulating Materials and Structures

Insulation is the most important element of all high-voltage devices, including EDT installations. It determines their lifetime and operation reliability. At present dielectrics of more than 400 types that differ significantly by complex of their characteristics (dielectric, mechanical, thermal, etc.) are used as solid dielectrics in electrical engineering and power engineering. The requirements for working gradients and operation reliability of insulating designs are especially high in EDT. This primarily concerns high-voltage insulation of drilling heads and drilling implements—devices for impulse transfer to the well bottom. It must satisfy the following main requirements:

(1) Reliable operation with a great number (up to 10^8) of voltage impulses and high (up to 20 kV/mm) working electric field gradients,
(2) Chemical resistance to the influence of the surrounding liquid environment (flushing-out liquids) which can be aggressive,
(3) Mechanical resistance to hydroshocks at breakdown of liquid environment and rock at the well bottom,
(4) Stability to shock loading of drilling implements during their cyclic motion in the well,
(5) Stability to abrasive wear in high-speed flow of the flushing-out liquid with drilling sludge.

Even without mechanical and chemical actions on the insulation, long-term exposure to an electric field gradually deteriorates its main characteristics—its electric aging is observed leading to insulation failure. Aging is caused by many processes: accumulation and redistribution of space charges; formation, evolution, and accumulation of submicron and micron cracks; chemical transformations; ionization processes in gas inclusions; and other destructive processes near structural defects. The intensity of these processes changes continuously in time and depends on a great

V. Y. Ushakov et al., *Electro-discharge Technology for Drilling Wells and Concrete Destruction*, https://doi.org/10.1007/978-3-030-04591-3_6

number of factors, first of all, on the electric field stress. Depending on the electric field stress and type, any of the above-indicated processes can dominate in electric aging. The development of engineering methods for calculation, design, and prediction of reliability of insulation operation under such adverse conditions are very complicated problems. To design insulating systems/structures, the developers are forced to increase the electrical strength safety margin to compensate for limited knowledge of action of different factors and lack of reliable methods for extrapolating results of laboratory investigations toward exploitation (operating) regimes. At present, a theoretical description of this complex multi-factor process is an unsolvable problem. The main method for determining the electrical strength of insulating materials under long-term exposure to voltage (electric aging) and short-term action of breakdown remains experimental.

At the very beginning of EDT development, no systematic studies of the electrical strength of solid dielectrics exposed to repeated action of voltage impulses were available. There were practically no data on reasons for aging macroscopically homogeneous monolithic polymeric insulation with large thickness. The above-indicated characteristics were investigated in more detail for combined (paper-oil, oil-film, and paper-oil-film) insulation for which the main factors of electric aging are partial discharges. Of particular interest for the EDT is a monolithic solid insulation with thickness of 10 mm and larger upon long-term exposure to voltage impulses of microsecond duration.

The need for knowledge of the characteristics of multi-impulse strength of insulation and of the nature of its electric aging under long-term exposure to voltage impulses has significantly increased in connection with the development and creation of high-voltage ED installations with electrical strength parameters close to "limiting" ones that additionally possess high reliability and strength. Besides, the development of unique insulating designs unavailable commercially was required.

Investigations of insulation of high-voltage technological installations started at the HVRI in 1963. They were performed in four directions: (a) search (choice) of insulation materials best satisfying the EDT requirements, (b) development of the production technology of large-size polyethylene structures for high-voltage (hundreds of kilovolts) insulators, (c) investigation of mechanisms of aging and failures of polymeric monolithic insulation exposed to voltage impulses, and (d) development of screening tests for insulators. Preliminary rejection of insulating structures (first of all, bushings and centering insulators of a drilling implement and the main insulators of drilling heads) is aimed at reducing the time of lowering/lifting of drilling implements and substitution of expensive insulating designs. Results of these studies obtained in 1987–1988 were generalized in monograph [1].

Based on the literature data and investigations of fifteen polymeric dielectrics performed at the HVRI, low-density polyethylene (LDPE) was chosen as a material for insulators of ED drilling implements.

6.2 Single-Impulse and Multi-impulse Strength of Insulating Materials and Structures in EDT

6.2.1 Methods of Sample Preparation and Testing

Coaxial insulators are used in a drilling implement for ED drilling of rocks. A photograph of insulators is shown in Fig. 6.1.

The coaxial insulators with smaller thickness and volume are also used in laboratory studies. The external view of the examined insulator sample is shown in Fig. 6.2.

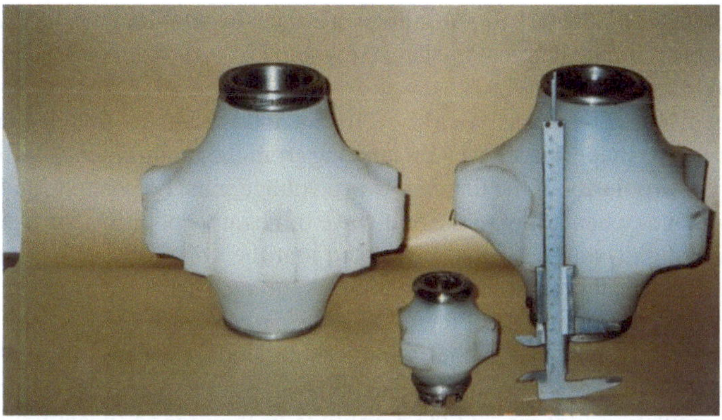

Fig. 6.1 External view of insulators of a drilling implement

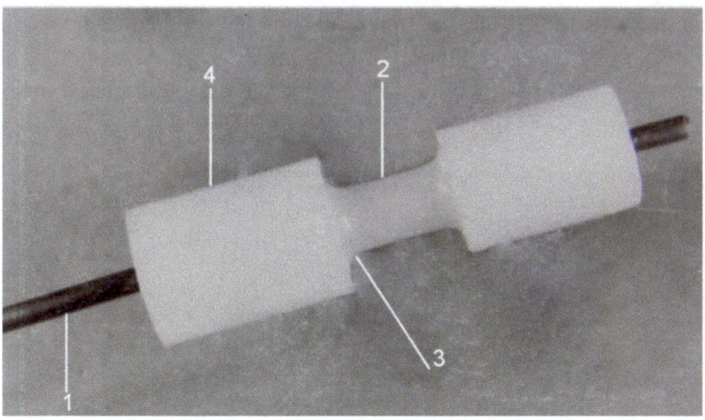

Fig. 6.2 Examined sample comprising internal high-voltage electrode 1, operating part 2, transitive region 3, and outer edge 4

The technology of re-processing of granulated polyethylene into an insulation design developed at the HVRI allowed samples to be manufactured without additional machining.

To eliminate partial discharges from the earthed electrode, a semiconducting layer is always deposited on the operating sample surface on which an earthed metal foil electrode is put. The semiconducting mixture with specific surface resistance $\rho = 20\text{–}50\ \Omega$ cm was developed together with the method of its deposition on the sample. The high-voltage electrode was placed into a press form in the course of sample preparation from granulated polyethylene before its melting.

To study the statistics of failures (breakdowns) of polymeric insulation samples during single- and multi-impulse tests, a high-voltage high-frequency test facility was specially developed whose main part was the VIG with a nominal voltage of 500 kV, capacitance per discharge of 0.00625 μF, and impulse repetition frequency up to 30 impulses/s. From 1 to 50 samples were tested simultaneously. The voltage impulse shape remained almost unchanged when the capacitive load increased from 9 to 450 pF, Fig. 6.3.

The impulse rise time was $\tau = 0.2 \times 10^{-6}$ s and the attenuation decrement was 1.8. The impulse shape was chosen proceeding from the requirements of ED drilling technology: the polarity of the first half-wave voltage must be positive, and the zero-crossing voltage (when going to the half wave of negative polarity), on average, was 0.3 of the first half-wave. It should be noted that the voltage waveform influenced significantly the insulation lifetime (the total resource). This was confirmed by experiments in which the amplitude of the first half-wave remained constant (its influence is a well-known fact), whereas the impulse front slope, the rate of impulse decay, and the amplitude and slope of the second half-wave were varied. The greatest effect on the PE insulation lifetime had the amplitude of the second (negative) half-wave and the front slope of the first half-wave. For example, an increase in the second half-wave amplitude by a factor of 2.83 reduced the number of average of impulses before breakdown by more than 100 times, and an increase in the impulse front slope by a factor of 8.5 caused the reduction of the number of impulses before breakdown by a factor of 2.4. Changes of other impulse parameters influenced insignificantly

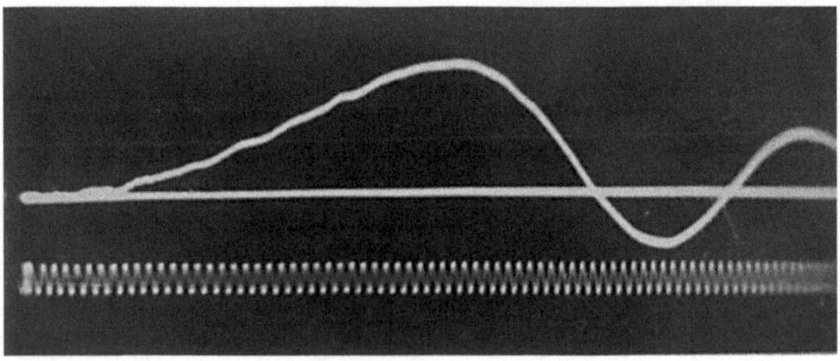

Fig. 6.3 Waveform of the test voltage impulse

on the number of impulses before breakdown in the examined range of variations of the impulse parameters characteristic for ED drilling.

In the well known methods for accelerating insulation lifetime tests, the acceleration parameters are chosen taking into account operating regimes of insulation. If the proportionality between the ratio of the testing to operating stresses and the ratio of the testing to operating lifetimes is reached, the acceleration is linear in character.

The electric field stress and the voltage impulse repetition frequency are most suitable for the acceleration parameters of insulation of high-voltage impulse installations. The frequency acceleration method is not widely used in practice of testing the insulation of electrotechnical equipment due to complexity of the development of sources of voltage impulses with amplitudes of several hundreds of kilovolts and high impulse repetition frequencies.

Below attention is focused on different regimes of test acceleration using the field stresses exceeding the operating field stress. This is important not only for establishing statistical regularities of polymeric insulation failure, but also for searching optimal conditions for rejection of insulating designs. In this case, it is important that the mechanisms of accelerated aging during tests were most identical to aging mechanisms during exploitation. Otherwise, the extrapolation of the accelerated tests to the operating conditions is not justified at all. A frequency of 30 impulses/s approximately corresponding to the maximal operating frequency acceptable for ED drilling (24 impulses/s) was used for sample testing.

The maximal and average magnitudes of the electric field stress were calculated from the classical formulas for coaxial systems. The error was determined by the accuracy of measuring U and the sample geometry and totally did not exceed 6.6%.

Test samples were prepared by melting of granulated polyethylene in vacuum at a temperature of 140 °C, pressing of melted polyethylene, and its subsequent cooling. Heating and cooling regimes were chosen based on the results of special studies. The applied technology of treatment of granulated polyethylene and the specially developed compression mould allowed 10 samples to be prepared simultaneously without additional machining.

6.2.2 Statistical Processing of Results

The spread of the results of measuring the electrical strength of dielectrics and the number of impulses before breakdown depend on the nature of the material, electrode design, voltage type, etc. For dielectrics with internal defects and inhomogeneous internal structure to which polyethylene insulation belongs, the electrical strength depends on the strength of the weakest link (weak spot) in the dielectric volume [2]. The spread of breakdown voltages and times before breakdown, caused by these reasons, does not obey a normal law [3–5]. It can be described by a normally logarithmic distribution [4] or one of the extreme distributions [5]. The theoretical approach to the choice of the law of breakdown voltage distribution based on the "weak link" theory assumes the application of one of the extreme distributions of minimal quantities.

The extreme distribution of minimal quantities of the third type or the Weibull distribution [5, 6] was most suitable. The mathematical expression of this distribution for short-term (single-pulse) electrical strength has the form

$$P(E) = 1 - \exp\left[-(E_i/E_e)^a\right], \tag{6.1}$$

where $P(E)$ is the probability of dielectric breakdown under application of the short-term voltage, E_i is the current value of the breakdown electric field stress, E_e is the breakdown field strength for $P = 1 - e^{-1} = 0.632$, and a is the parameter determined experimentally. Therefore, it is necessary to know the parameters of the chosen distribution. Such parameters for the Weibull distribution given by Eq. (6.1) are E_e and a.

The dielectric breakdown (failure) occurs upon long-term exposure to voltage at the voltage lower than the short-term breakdown voltage. It is considered that the breakdown probability of the region electrically weakened upon long-term exposure to voltage is higher than of all other regions [2]. Among these weakened regions are air inclusions, regions with local increase of the electric field stress in the sample volume, etc. The characteristic feature of the failure upon long-term exposure to voltage is a considerable spread of times before breakdown. In this case, the average lifetime of the sample is conditional, and the characteristic of the behavior of sample failures is incomplete. The time distribution of failures is described by the Weibull distribution of the form [2, 6–8]

$$P(t) = 1 - \exp\left[-(t_i/t_e)^b\right], \tag{6.2}$$

where $P(t)$ is the probability of sample failure during time t, t_i is the current time, t_e is the time before failure for $P = 1 - e^{-1} = 0.632$, and b is the parameter determined experimentally. Here t_e and b are the parameters of the Weibull distribution.

Thus, the electrical strength and failures upon long-term exposure to voltage can be described by the same probability distribution law. This makes it possible to describe more complex phenomena from the probabilistic viewpoint. For example, this is the case for failures for one batch of samples at different voltage levels, "lifetime curve," etc.

The dependence of the probability of sample failures upon long-term exposure to voltage for different magnitudes of the electric field stress is described by the Weibull distribution of the form

$$P(E, t) = 1 - \exp\left[-(E_i/E_e)^a \cdot \left(\frac{t_i}{t_e}\right)^b\right], \tag{6.3}$$

$$\text{where } t_e^b \cdot E_e^a = \text{const.} \tag{6.4}$$

Expression (6.4) can be written as follows:

$$t_e \cdot E_e^n = \text{const}, \tag{6.5}$$

where $n = a/b$ describes the rate of insulation aging. This is the well-known equation of the "lifetime curve." Other E and t values for the preset probability P can be included into Eq. (6.5) instead of E_e and t_e. In the $\log t - \log E$ coordinates, the "lifetime curve" given by Eq. (6.5) represents a straight line for testing field stresses $E_e = 13.8–180$ kV/mm used in experiments described here.

To analyze the reliability of insulation operation based on the results of sample testing, the probability density distribution or the sample failure rate must be known. Disadvantages of application of the failure rate for an analysis were demonstrated in [4]. Therefore, it is expedient to use the probability density distribution of sample failures $F(t)$, Fig. 6.4. This distribution indicates the number of failures in each time interval. In reliability theory, it is considered that the first region of enhanced probability density corresponds to failures during operation, and the stabilized density corresponds to a random failure or a failure from wear.

6.2.3 *Multi-impulse Electrical Strength of Polyethylene Insulation*

Failures of both laboratory samples and actual electroinsulating designs in the Weibull plot have several—from one to three—rectilinear fragments. Their presence indicates different natures of failures. In most cases, two rectilinear fragments are observed in the Weibull plots, Fig. 6.5.

The first fragment (initial failures) is caused by technological imperfections (defects) in the insulation volume, and the coordinates of the bend point of the Weibull plot of failure distribution determine precisely the relative number (the fail-

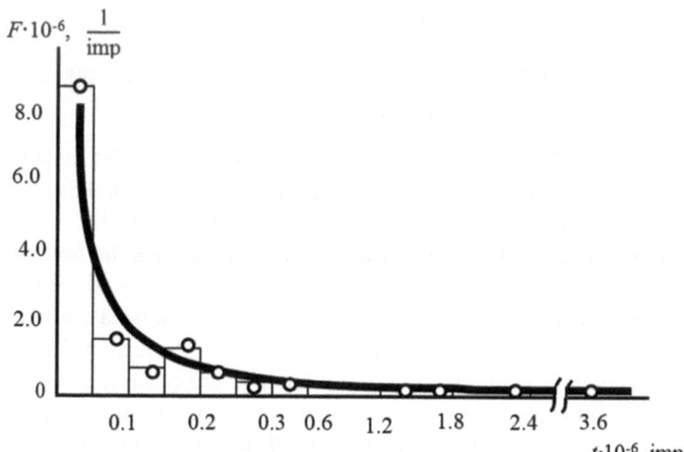

Fig. 6.4 Density of LDPE sample failures for $E_{test} = 50$ kV/mm

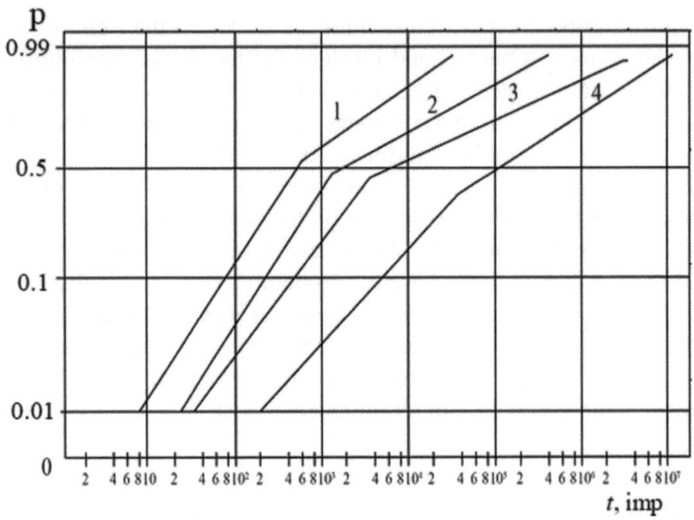

Fig. 6.5 Distribution of sample failures in the Weibull plot for $E_{test} = 100$ (curve 1), 83.5 (curve 2), 66.4 (curve 3), and 50.0 kV/mm (curve 4)

ure probability) of the defective samples and the number of impulses causing their failure.

This effect is retained when all testing conditions change, including:

(1) Materials,
(2) Polyethylene types (polyethylene of 6 types was tested),
(3) Technology of polyethylene sample production and regimes of temperature annealing,
(4) Change of the working volume of insulation in the range 0.126–140.000 mm^3,
(5) Change of testing electric field stress in the range 13.5–180.0 kV/mm,
(6) Change of the impulse repetition frequency in the range 1–1000 impulses/s,
(7) Waveforms and parameters of voltage impulses.

The efficiency of the suggested method for estimation of the insulating design quality based on the bend point of the Weibull plot was confirmed by investigations performed by the nonparametric method of quality estimation [9].

This method for rejecting electroinsulating designs allows the number of defect designs and the time required for their rejection to be evaluated without *a priori* choice of a theoretical failure distribution function and a physical model.

The dependences of the "lifetime" (Eq. 6.2), the "lifetime curve" (Eq. 6.3), and hence, the parameters a, b, and E_e, t_e of the Weibull distribution and of the coefficient n for the conditions listed in items 1–5 and for the bend points of the Weibull plot were obtained along with restrictions on the applicability of the calculation formulas based on a great volume of experimental data and analysis of the results obtained with confidence intervals of 0.9. In particular, the number of impulses before the bend points $t_{bend\ point}$ for different test field stresses E were described by the "lifetime curve" equation

$$t_{bend\ point} = (t_{bend\ point1} \cdot E_1^n) \cdot E^{-n}, \tag{6.6}$$

where $t_{bend\ point}$ and $t_{bend\ point\ 1}$ denote the number of impulses before the bend point at E and E_1, respectively, and n is the experimental coefficient. Changes of the probability of occurrence of the bend point $t_{bend\ point}$ depending on the test field stress are described by linear equation:

$$P_{bend\ point} = C + dE. \tag{6.7}$$

The coefficients in formulas (6.6) and (6.7) for the coaxial LDPE polyethylene samples are $n = 7.16$, $C = 0.221$, and $d = 0.0024$.

From Eqs. (6.6) and (6.7) it follows that an increase in the test field stress E causes the decrease of the number of withstand impulses and an increase in the relative and absolute number of samples failed before the bend point. This is due to additional defects responsible for failures in the first segment of the Weibull distribution formed with increasing E that at lower E values do not yield "technological defects." To reject all technologically defective designs before the bend point in the plot of the Weibull distribution, a certain number of impulses $t_{bend\ point}$ are required which, in turn, depends on the amplitude of the test voltage impulses. The best variant is to carry out rejection at the operating voltage not to reduce the lifetime of the remaining high-quality designs. However, realization of such approach requires a very great number of rejection voltage impulses (10^4–10^5 impulses). Hence, the question arises: How can we choose the test field stress for accelerated tests?

To determine optimal conditions for the rejection tests, the influence of preliminary exposure to voltage on multi-impulse and one-impulse electrical strengths of polyethylene insulation was investigated. In the English literature, this procedure is called "prestressing."

6.2.4 Influence of the Preliminary Exposure to Voltage Impulses on the Single-Impulse Electrical Strength of Polyethylene

The influence of the electric field on the subsequent behavior of dielectrics is manifested either through a decrease in the time before breakdown with increasing field stress (displayed by the "lifetime curve"), or through a decrease in the short-term electrical strength upon preliminary exposure to voltage (direct, alternating, or impulse). The short-term electrical strength (for alternating and direct voltages) or the single-impulse electrical strength (for impulse voltage) is the main characteristic of insulation.

To choose the optimal voltage and the number of test impulses for rejection of insulating designs, the influence of preliminary single-impulse aging on the change of the impulse breakdown voltage was investigated. To study preliminary aging, at each voltage level corresponding to (0.9–0.6)E_{br}, a dependence of E_{br} on the number of applied impulses was determined. In this case, E_{br} for alternating voltage decreased by 45–60% when the number of aging impulses increased to 50 and the test ("aging") field stress increased to 0.9E_{br}. The test field stress was determined at half maximum of the breakdown field stress E_0 for the batch of samples that have not been preliminary tested. Magnitudes of the test field stress were chosen from the range 0.057E_0–0.453E_0.

The maximal number of test impulses was chosen so that to exceed the number of impulses required for rejection. The intermediate numbers of test impulses were chosen arbitrarily. The number of samples was such that after preliminary tests no less than 20 samples remained which then were broken down on the front of a single impulse.

The method of statistical data processing suggested in [4] was used, according to which samples rejected during preliminary tests were also taken into account when determining the breakdown probability. They were included in the breakdown probability given by the formula

$$P = \frac{m + m_i}{N + 1}, \tag{6.8}$$

where m is the number of samples rejected during preliminary aging, m_i is the current number of samples subjected to single-pulse breakdown, and N is the number of samples in the batch.

Figure 6.6 shows the distributions of the maximal breakdown electric field stress E_{br} in the Weibull plot for $E_{test} = 25$ kV/mm with all other testing conditions remaining the same.

The distribution function for E_{br} of samples that were not preliminarily aged ($t_{test} = 0$ impulses) has two rectilinear segments that can be caused by manifestation of technological defects.

Figure 6.7 shows the dependences of the parameter E_e of the Weibull distribution on the number of test impulses during preliminary aging. For all test field stresses the maximal breakdown field stress was observed at $t_{test} = t_{opt}$ that was lower than the initial E_e magnitude and varied in a complex manner as a function of E_{test} and t_{test}.

Figure 6.8 shows the dependences of the Weibull distribution parameters E_e and a on the test electric field stress for $t_{test} = t_{opt}$. It can be seen that the parameter E_e has a maximum, whereas the parameter a has a minimum for the same E_{test} magnitude.

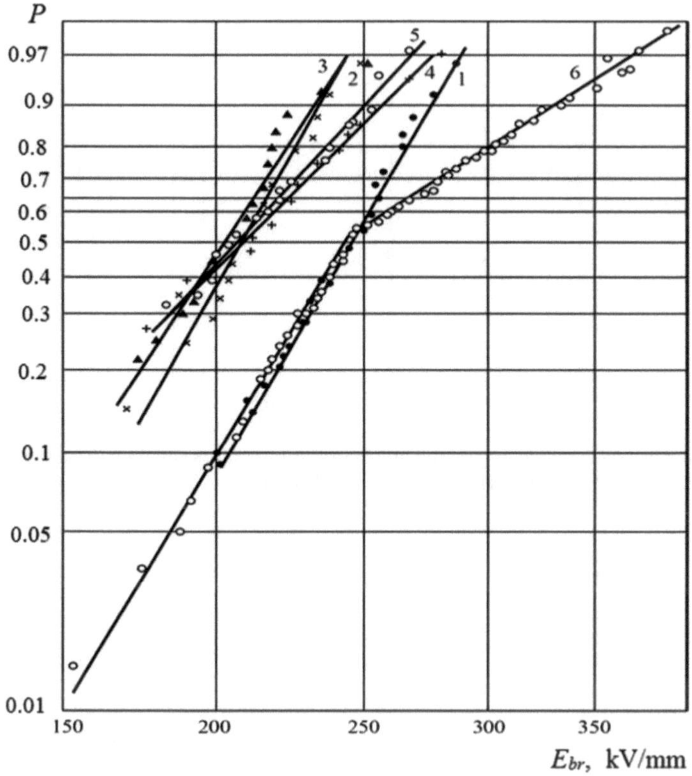

Fig. 6.6 Distribution of the breakdown field stress in the Weibull plot for $E_{test} = 25$ kV/mm and the number of test impulses equal to 1000 (curve 1), 50,000 (curve 2), 400,000 (curve 3), 1000,000 (curve 4), and 3200,000 (curve 5). Here curve 6 is for 0 impulses

Hence, the single-pulse breakdown field stress E_{br} of polyethylene samples depends on the stress amplitude during preliminary aging E_{test}. This dependence has a maximum in the range $E_{test} \leq 0.3E_0$ upon exposure to the optimal number of aging impulses t_{opt}.

These results allow us to assume that the single-pulse electrical strength of polymeric insulation during aging depends on a combination of the test field stress and the number of voltage impulses even after rejection of technological defects.

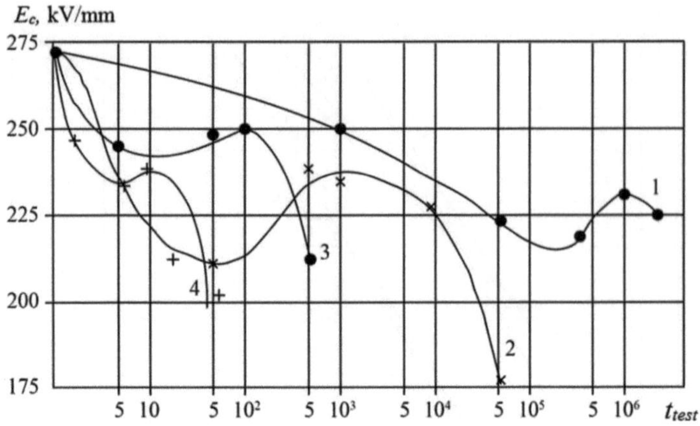

Fig. 6.7 Change of E_C depending on the number of test impulses equal to 25 (curve 1), 50 (curve 2), 80 (curve 3), and 111 (curve 4). Here $E_{test} = 13.8$ kV/mm

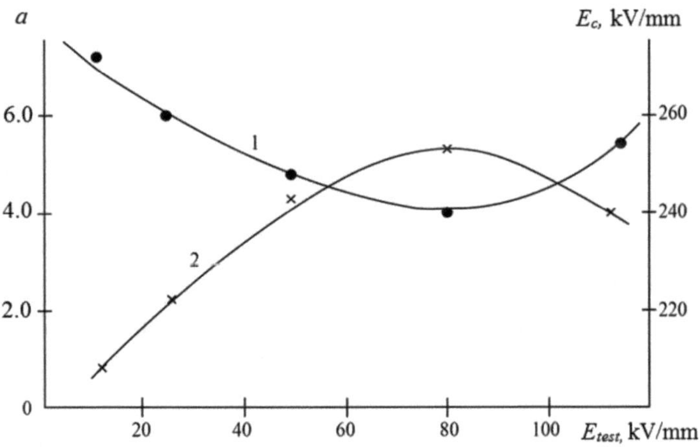

Fig. 6.8 Dependence of the parameters of the Weibull distribution on E_{test} for $t_{test} = t_{opt}$. Here curve 1 is for a and curve 2 is for E_e: short-term electrical stress (for variable and direct voltage) or single-impulse electrical stress (for impulse voltage)

6.2.5 *Influence of the Preliminary Exposure to Voltage Impulses on the Multi-impulse Electrical Strength of Polyethylene*

A decrease in the breakdown field gradients though indicates the degree of dielectric aging, but cannot unambiguously predict its further behavior upon long-term exposure to reduced (operating) voltage. It is not yet possible to relate directly the decrease of the breakdown field gradients for a dielectric after aging time with its subsequent "lifetime" [5].

The influence of preliminary applied increased impulse voltage on the subsequent number of impulses before breakdown (“lifetime”) of polyethylene samples for the chosen operating impulse field stress E_{oper} was also investigated. Most samples were made of low-density polyethylene (LDPE). The procedures of sample preparation and multi-impulse testing were analogous to the above described. The maximal test field stresses at the internal electrode of the sample chosen for preliminary aging were $E_{test} = 66.4$, 88.3, and 100.0 kV/mm, and the maximal operating field stress was $E_{oper} = 50.0$ kV/mm. The maximal number of impulses for preliminary aging was chosen larger than the number of impulses necessary for sample rejection.

The results obtained were processed by the same technique that was used to estimate the decrease of the single-pulse electrical strength during preliminary aging [4]. It was limitedd to an analysis of the parameters b and t_e of the Weibull distribution.

Figure 6.9 shows the distributions of the failure probabilities for the operating field stress $E_{oper} = 50$ kV/mm depending on the number of impulses with amplitude creating the field stress $E_{test} = 83.6$ kV/mm. The dotted curves show the dependences for the test (curve 5) and operating (curve 6) field stresses without preliminary aging. For all other testing conditions, the character of changes of the failure distributions was analogous to that shown in Fig. 6.9.

After preliminary aging, two rectilinear segments can be seen in the dependence $P = f(t_{test})$. The occurrence of failures after preliminary aging for their relatively low probability (the first linear fragment) was caused by a more intensive character of insulation aging and involving in aging of defects that were not recorded for the operating field stress.

Joint consideration of the parameters of the second fragment t_{e2} and b_2 allows the conclusion to be drawn that when E_{test} becomes greater than $E_{test} = 1.33E_{oper}$, the

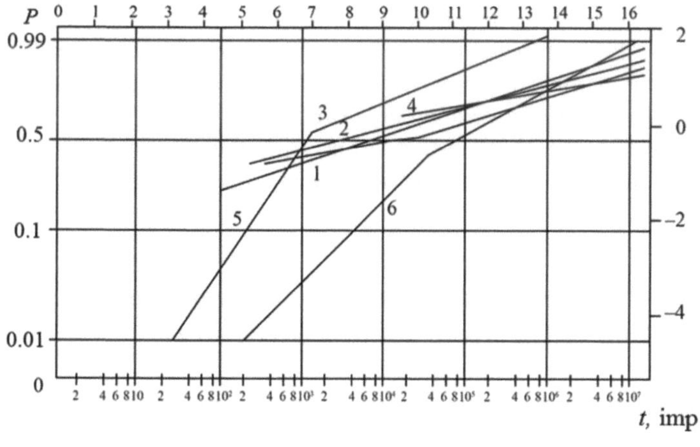

Fig. 6.9 Distributions of failures for the Weibull plot of preliminary tested samples at $E_{test} =$ 83.6 kV/mm and the number of test impulses equal to 500 (curve 1), 1000 (curve 2), 2000 (curve 3), 5000 (curve 4), and 0 (curve 5). Here curve 6 is for $E_{oper} = 50$ kV/mm

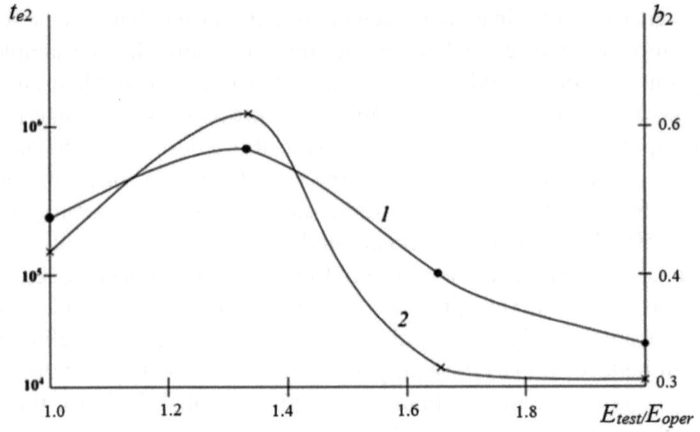

Fig. 6.10 Change of the Weibull parameters t_{e2} and b_2 of the dependence on E_{test}/E_{oper}. Here curve 1 is for t_{e2} and curve 2 is for b_2

characteristics of multi-impulse strength of polyethylene samples for operating field stress are deteriorated (Fig. 6.10).

If we compare the change of the parameters E_e and a (Fig. 6.8) with t_{e2} and b_2 as functions of E_{test} (Figs. 6.8 and 6.10) for $t_{test} = t_{opt}$, it turns out that E_e and t_{e2} have maxima, whereas a and b_2 have opposite character of changing. From this it follows that after preliminary aging, it is impossible to judge unambiguously on the change of the multi-impulse strength based on the change of E_{br}.

The preliminary tests of polyethylene samples with increased voltage must be performed for each stress amplitude and strictly assigned number of impulses corresponding to complete rejection of the technological defect.

An increase in the test field stress amplitude above $1.33E_{oper}$ leads to a sharp reduction of the number of impulses before failure and of the parameters of the Weibull distribution at the operating field stress.

To increase the reliability of insulation operation, the polyethylene samples should be rejected preliminary at field stress amplitudes $E_{test} \leq 1.33E_{oper}$.

6.3 Basic Principles for Designing and Preliminary Rejection of Electroinsulating Structures

In EDT it is necessary to deal with a system of three electroinsulating parts exposed to an impulse electric field. In drilling installations, they are "rock–insulating and flushing-out liquid–solid insulation of a drilling implement." In other EDT applications, they are "destructible/treatable solid body—insulating/working liquid–solid insulation of the working tool." In this system, matching of the voltage-time characteristics of the two-component systems "rock–liquid insulation" and "liquid insulation–solid insulation" is required. Conditions of their operation and requirements imposed on them differ significantly. If the solid undergoing treatment/destruction must be broken down upon exposure to each impulse, the solid insulation of the drilling implement (generally actuator) should withstand millions of voltage impulses without breakdown and surface flashover.

As indicated above, the insulation of a drilling implement (DI) includes bushings and insulator of a drilling head. The above-formulated requirements can be met first, by the choice of the electric field configuration: sharply inhomogeneous at the well bottom and slightly inhomogeneous (coaxial) in the transmission line (drilling tubes). Second, the solid insulation can be matched by adjusting the corresponding drilling tube diameter.

The matching condition can be written in the following form:

$$U_{rock} < U_{br.l} < U_{br.sol}. \tag{6.9}$$

Since the breakdown of liquid insulation is undesirable but does lead to the DI failure, the condition $U_{br.l} < U_{br.sol}$ is acceptable. This condition does not contradict with the above-considered phenomenon of excess of the electrical strength of liquids over that of solid dielectrics, because the breakdown of the liquid dielectric takes place in the region of the drilling head where the electric field is sharply inhomogeneous. The processes of surface flashover of solid dielectrics must also be taken into account. In this regard, condition (6.9) is written as

$$U_{rock} < U_{br.l} < U_{fl}, \tag{6.10}$$

where U_{fl} is the surface flashover voltage of the solid insulation of the drilling implement. In this regard, the question arises on the influence of the insulator material and shape, the location of apertures in it (for circulation of the flushing-out liquid), and the liquid flow rate.

From all insulating materials (ftoroplast-4, polymethylmethacrylate, textolite, polystyrene, epoxy compound, zircon-based ceramics, and low- and high-density polyethylene) acceptable in principle for EDT and investigated thereupon, the maximum flashover voltage in the range of examined voltage exposure times $(0.4–6.0) \times 10^{-6}$ s and at the least flashover probability are characteristic for high- and low-density polyethylene. It should be noted that the flashover voltage, like the breakdown voltage, increases with decreasing impulse duration. Tests of DI insulators during drilling have demonstrated that insulators made of epoxy compound with filler from polymethylmethacrylate or polystyrene possess insufficient mechanical strength leading to their destruction during lowering-lifting operations or under breakdown. The insulation flashover voltage increases with the flow rate of the insulation liquid (transformer oil) for all insulator shapes. In this case, conic or concave insulators with peripheral apertures for liquid circulation have the maximal flashover voltage.

Insulators of drilling implements made of low-density polyethylene were equipped with conic surface. The apertures for liquid circulation were located on the insulator periphery. The external view of the insulator is shown in Fig. 6.1. Table 6.1 gives the field stresses for the indicated drilling implements corresponding to the maximal voltages withstand by insulators without flashover.

The dimensions of the DI insulators cannot be changed arbitrarily—they are determined by the internal diameters of the column of drilling tubes and high-voltage electrical wires. The sizes of the electrical wire should be determined by the relationship between the diameters of the coaxial system at which the insulator "lifetime" reached its maximum (Table 6.1).

By the present time, several DI types have been developed and passed or are passing laboratory and field tests. The reliability of their operation depends on the reliability of operation of their individual parts: high-voltage input, electric conductor, column of drilling tubes, drilling head, solid insulation, etc.

The relative contribution of some auxiliary operations to the total DI operating time (about 287 h) is given in Table 6.2. The results were obtained in the first years of field tests.

These relationships change with further development of ED drilling technology, but the main reason of unproductive labor expenditures is the solid insulation breakdown. In this regard, the problem of improving DI insulating designs remains urgent.

From Table 6.1 it follows that U_{rock}, that is, the voltage used for drilling is always less than U_{fl} for the given S. With sufficient degree of confidence one can speak

Table 6.1 Structural and electric characteristics of drilling implements

EIDI type	D/d	Δ, mm	S, mm	U_{fl}, kV	U_{rock}, kV	E_{max}, kV/mm	E_{av}, kV/mm
BI-89-15	3.2	27.5	15	385	220	15.1	8.0
BI-108-20	2.78	32.0	20	445	240	13.0	7.5
BI-108-15	2.78	32.0	15	445	220	12.0	6.9
BI-108-20 V	2.38	29.0	20	406	240	13.2	8.3
BI-108-30	2.78	32.0	30	445	280	15.2	8.7
BI-127-50	3.33	42.0	50	537	360	16.6	8.6
BI-219-70	2.72	64.0	70	704	420	10.7	6.6
BI-325-120	2.87	101.0	120	798	486	8.6	5.0

Table 6.2 Time constituents of electroimpulse drilling

Drilling, %	Accidents and failures, %			Down time, %
	Insulator breakdown	Malfunctions of the VIG	Other failures and accidents	
35	25	2	25	13

about the absence of solid insulation destruction due to surface flashover. Separate discharges observed on the insulator surface do not lead to insulator failure.

Judging by the data presented in Table 6.1, the thickness Δ of the insulating layer of insulators was 27.5–101 mm. Fabrication of high-quality insulation of such thickness from polyethylene presents considerable difficulties. The first batches of DI insulators were fabricated from block high-pressure polyethylene. These insulators demonstrated low operation reliability. This is not surprising, because cavities of large sizes—up to several millimeters—were observed in the polyethylene volume.

The LDPE insulators fabricated in the industrial plant by extrusion with fast cooling had the same disadvantages as block polyethylene ones. Therefore, the technology of centering insulator fabrication from granulated polyethylene already mentioned above has been developed at the HV RI. A total of 112 insulators for drilling implement BI108-30, Table 6.1, were fabricated using this technology. All insulators were preliminary rejected at the maximal stress on the internal electrode E_{max} = 16.0 kV/mm that was close to the operating stress (E_{max} = 15.2 kV/mm). The insulators were exposed to 5 × 10^4 impulses corresponding to the bend point of the Weibull curve. The first failure during rejection test occurred after 60 impulses. During rejection tests, 52% of insulators with technological defects failed. In the course of drilling by the BI-108-30 drilling implement and in the subsequent insulator tests, 15.2 × 10^6 pulses were applied to the insulators. The first failure at the operating voltage after rejection tests was observed after 1.7 × 10^5 impulses, and the average time between failures was 5 × 10^5 impulses. In this case, no breakdown was observed for 45% of samples subjected to preliminary rejection. This result can be considered satisfactory for the initial stages of EDT development for well drilling.

The next stage of modernization of the DI design was initiated by results of drilling at the Leninogorsk test site (nowadays in the Kazakhstan). The DI of increased overall dimensions with drilling head diameter D_{tir} = 200 mm, interelectrode distance S = 50 mm, drilling tube column diameter D_{DT} = 127 mm, and operating voltage gradients given in Table 6.1 was developed for prospecting in Kolyma (Russia). Despite the increased diameter of the drilling tube column, the developers succeeded in increasing the maximum field stress of insulation up to E_{max} = 16.6 kV/mm. A batch of LDPE insulators was preliminary tested upon exposure to 5 × 10^4 impulses of operating voltage. Drilling was performed in permafrost coarse skeleton deposits at air temperature of −53 °C close to the LDPE vitrification temperature. No more than 7 insulators were simultaneously operating in the drilling implement because the drilling depth did not exceed 21 m. Altogether, 72 m of wells were drilled. The insulators withstand 2.66 × 10^5 impulses. In this case, 3 insulators failed because of fragile splitting when the drilling implement hit the well bottom.

The applicability of the given approach to the choice of optimal rejection test regimes for actual insulating designs was tested using 650 sections of the RK-75-4-11 cable 1 m long disregarding cutting. The chosen rejection conditions were typical for cable operation in the regime of transfer of vibrational impulses of microsecond duration with repetition frequency of 1–2 impulses/s. The insulation "lifetime" was set equal to 10^5 pulses with reliability of 0.95 at maximal field stress of 50 kV/mm. The number of cable sections for individual tests was no less than 20. Based on the test results, the rejection regime with the parameters $E_{test} = 70$ kV/mm and $t_{test} = 10^4$ impulses was adopted as optimal. The control cable tests demonstrated that for the cable batch rejected in the optimal regime, the operating time to the first failure increased by 1.5 orders of magnitude in comparison with the non-rejected cable batch, and the probability of failure for 10^5 impulses decreased by 10 times, Fig. 6.11.

Subsequent modernization of the technology based on the results of test drilling and laboratory investigations allowed the maximal electric field stress (E_{max} = 16.6 kV/mm) and the number of impulses before failure to be increased. In this case, the average number of impulses before failure increased by more than 7 times.

An empirical method of calculation of the number of impulses before breakdown of solid insulation, the available laboratory results obtained, as a rule, for high gradients and small volumes of insulation material must be primarily extrapolated to the operating conditions. For this purpose, the well-known and well-proven equation of the "lifetime" curve (Eq. 6.8) was used. It can be written in the form where n is the parameter characterizing the degree of aging and t_0 is the number of impulses before breakdown at E_0.

$$t = (t_0 \cdot E_0^n) \cdot E_{oper}^{-n}, \tag{6.11}$$

Equation (6.11) allows one to predict the number of impulses before insulation failure at minimum stress levels E_{oper} = (10–15) kV/mm with high degree of reli-

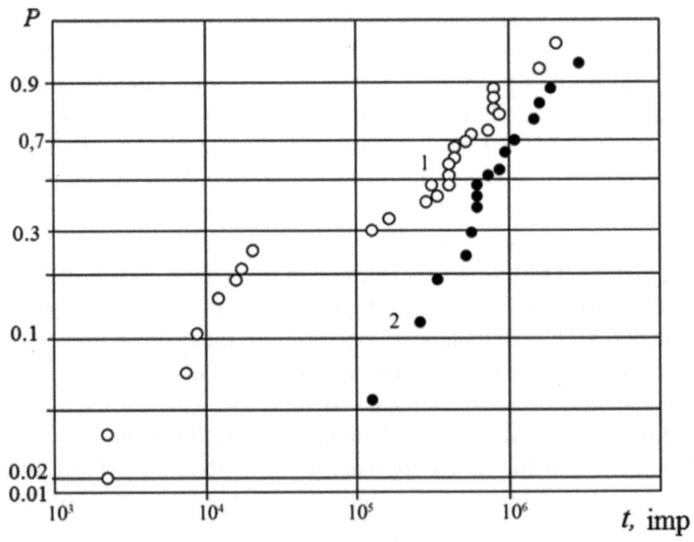

Fig. 6.11 Distribution function for the number of impulses before breakdown of RK-75-4-11cable sections at $E_{test} = 50$ kV/mm without rejection (curve 1) and with rejection at $E_{test} = 70$ kV/mm and $t_{test} = 10^4$ impulses

ability. Further reduction of E_{oper} in calculations leads to considerable insulation safety margin. Knowing t_{e0} at E_0, the number of impulses before breakdown can be determined for the Weibull parameter t_e at E_{oper} given by Eq. (6.2). We are interested in the "lifetime" at a preset level of reliability, for example, at $Q = 0.99$. Then the probability of breakdown should not exceed $P = 1 - Q = 0.01$. The number of impulses before breakdown for such probability can be determined from Eq. (6.2). The dependence of *b* on the electric field stress for insulation required for this has the form

$$b(E) = 4.8 + 0.05E_{test}, \tag{6.12}$$

where E_{test} is in kV/mm. This expression is in agreement with modern notion of the behavior of the variance as a function of *E*. Thus, the insulation "lifetime" can be calculated with preset reliability from the three equations presented above based on the initial experimental data.

Since the dimensions of the actual insulation designs are greater than those of the laboratory samples, the effect of dielectric volume—the reduction of the electrical stress and the "lifetime"—must be taken into account for increasing volume of insulation. Results of investigations of this effect for direct, alternate, and impulse voltages allowed one to write down the following expression:

$$t(v) = (t_0 \cdot v_0^{c_1})v_1^{c_1} = t_0\left(\frac{v_0}{v_1}\right)^{c_1}, \tag{6.13}$$

where c_1 is the coefficient characterizing the influence of change in the insulation volume, t_0 is the number of impulses before breakdown for the known volume. This expression is applicable for volumes changing by 6 orders of magnitude. The working volume of the insulator calculated by us was $v_1 = 6.6 \times 10^3$ cm^3, and the volume of experimental samples accepted as initial was $v_0 = 2$ cm^3, that is, the difference was approximately 3 orders of magnitude, which is quite acceptable. The coefficient c_1 was equal to 2.02 for the given operating conditions. Substituting t_e instead of t_0 into Eq. (6.13), we determine t_{e1} for a new volume v_1, and from Eqs. (6.2)–(6.12) we determine *t* for a preset level of reliability.

The parameter *b* in Eq. (6.2) changes as a function of the volume *v*; therefore, it is necessary to have a new *b* value for the new volume *v*. The dependence $b = f(v)$ can be represented in the form

$$b_1 = b_0 + k \cdot (\log v - \log v_0), \tag{6.14}$$

where b_0 is the parameter value for v_0 and *k* is a coefficient.

Equation (6.14) demonstrates that the variance decreases with increasing *v*. This does not contradict to the modern notion of the breakdown. In our conditions, Eq. (6.14) can be written as follows:

$$b_1 = 5.3 + 0.96 \cdot \log v. \tag{6.15}$$

Table 6.3 Experimental values of the parameters in formulas (6.13)–(6.16)

V_0, m^3	E_0, kV/mm	t_{e0}, impulses	n	c_1	b_0
2×10^{-6}	50	1.03×10^{10}	8.05	2.02	4.8

By this method, the insulator for the coaxial line of the BI-325-120 drilling implement was calculated. For example, if we take the operating voltage $U_{oper} = 500$ kV for drilling of granite with $S = 120$ mm, the maximal electric field stress in insulation of the coaxial system will be

$$E_{oper} = \mathrm{U}_{oper}/r \cdot \ln R/r = 8.78 \text{ kV/mm}.$$

Table 6.3 gives values of the parameters entering into Eqs. (6.2) and (6.11)–(6.15) and obtained experimentally for polyethylene samples.

Calculation of the "lifetime" for the parameter of Weibull distribution t_e from Eq. (6.11) for samples with volume $v_0 = 2.0$ cm^3 at $E_{oper} = 8.78$ kV/mm yields $t_{e1}(E) = 1.24 \times 10^{16}$ impulses. For the insulator of the BI-325-120 drilling implement with volume $v = 6.6 \times 10^3$ cm^3 at $E_{oper} = 8.78$ kV/mm, we obtain $t_{e2} = 1.1 \times 10^9$ impulses.

We have calculated one parameter of Weibull distribution—t_{e2}—which corresponds to the failure probability $P = 0.632$.

To ensure the required insulator "lifetime" (the number of impulses before failure), we set $P = 0.01$. Another parameter of distribution in Eq. (6.2)—b—depends on E_{oper} and v. Calculation from Eqs. (6.12) and (6.15) yields the value $b(E, v) = 3.17$. Then expression (6.2) after transformations assumes the form

$$t_i = 0.2343 \cdot t_{e2}, \tag{6.16}$$

and the number of impulses before breakdown at $P = 0.01$ will be $t_{0.01} = 2.58 \times 10^8$ impulses.

In the current stage of the development of EDT for drilling, this $t_{0.01}$ value allows several thousand meters of wells to be drilled without insulator failure.

It should be recognized that the reliability of the above calculations is not high, since their results have not yet been tested for long-term drilling.

The problem of insulation of the actuator in installations for the destruction of reinforced concrete products is considered in Sect. 8.4.3.

References

1. Koykov SN, Tsikin AN (1968) Electrical aging of solid dielectrics. Publishing house Energiya, Leningrad, 186 p (in Russian)
2. Artbauer J (1970) Elektrische Dauerfestigkeit und Kurzzeitfestigkeit. ETZ—A 91(6):326–331.9 (in German) 40

3. Ushakov VY (2004) Insulation of high-voltage equipment. Springer, Berlin, Heidelberg, 421 p
4. Oudin J-M, Rerollect J, Thevenon H (1968) Theorie statistique a claquage electrique. Rev Gn Elec 77(4) (in French)
5. Occhini E (1971) A statistical approach to the discussion of the dielectric strength in electric cables. IEEE Trans Power Appar Syst 90(6):2671–2682
6. Adam A, Vajov V (2000) Pat US 6,164,388 Electropulse method of holes boring and boring machine, 26.12.2000
7. Fabiani D, Simoni L (2005) Discussion on application of the Weibull distribution to electrical breakdown of insulating materials. IEEE Trans Dielectr Electr Insul 12(1):11–16
8. Kalyatsky II, Chepikov AT, Fortes YB, Pobyzhimov NF, Zykov VM, Ryabchikov SY, Simonov VY, Klein RE, Ishmukhametov ST, Peltsman SS, Vajov VF, Shishkin VS (1966) Certificate 971987. Drilling implement for drilling wells by electric pulse method with continuous cleaning of the bottomhole face from the products of rock destruction. Publ 02/25/1966 (in Russian)
9. Vajov VF, Zinoviev NT, Ushakov VY (2016) Electrodischarge technology for drilling wells and fracture of reinforced concrete products: monograph. In: Ushakov VY (ed). Publishing house of Tomsk Polytechnic University, 312 p (in Russian)

Chapter 7
Drilling of Wells

To drill wells, the main element of a drilling rig—a drill head—is designed as a system of potential (high-voltage) grounded electrodes uniformly distributed over the well bottom and nearly equally spaced, Fig. 7.1.

This design allows the EDT idea to be implemented, i.e., the superiority of the liquid medium in the electric strength for steeply sloping high-voltage impulses to be used for destruction of solids. The most important feature of this drilling rig is that it is not necessary to rotate it pressing to the surface of the destructible body, because

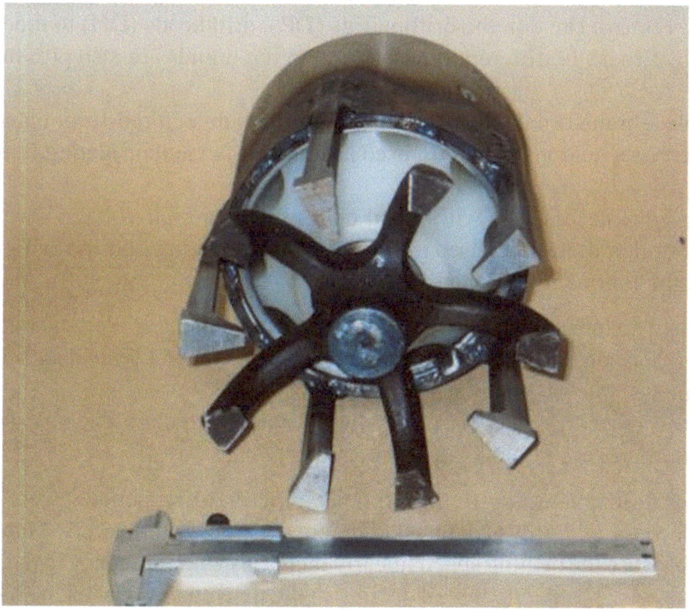

Fig. 7.1 Appearance of the drill head 150 mm in diameter

V. Y. Ushakov et al., *Electro-discharge Technology for Drilling Wells and Concrete Destruction*, https://doi.org/10.1007/978-3-030-04591-3_7

electric discharges are automatically spread over the gaps. In case of breakdown in one of the gaps (for example, in gap 1 of Fig. 5.3), the crater formed in the rock is filled with flushing-out liquid that leads to its electric hardening. The subsequent discharges occur successively in other gaps until the destruction of rock takes place in all gaps in the entire well bottom. After that the drilling rig moves to a new location in which all interelectrode gaps acquire approximately identical rock breakdown voltages. This cycle is repeated with the periodicity determined by the parameters of technological regime of the drilling rig.

The well bottom can have practically arbitrary shape, including round or rectangular one, by changing the geometrical parameters of the drill head (the shape and the number of electrodes and their spacing and mutual locations). The region of the well bottom can also be changed in wide limits (up to ten square meters and more).

The arrangement of electrodes only on the perimeter rather than over the entire well bottom area allows prospecting wells to be drilled with core sampling. This has been confirmed by laboratory and field tests.

7.1 Experience of ED Drilling in the Initial Stages of EDT Development

Table 7.1 generalizes main results of field tests of ED drilling rigs in the period till 1985 (most of these works were performed under the leadership of A. T. Chepikov and Yu. B. Fortes). Data on the drilling rigs (DR), drill heads (DH) to them, drilling speeds (V) of rocks of different types, and working liquids are also presented in the table.

* "Cable + transformer oil" denotes the wire manufactured from cable without external screen with transformer oil used as flushing-out and insulating liquids in the drilling rig.

** "Oil" denotes the DR with metal core used as a high-voltage wire and with transformer oil or diesel fuel used as insulating and working liquids. In Table 7.1 we have used the following designations:

- DR-44 and so on stand for drilling rigs 44 mm in diameter,
- DH-50-8 and so on stand for drill heads 50 mm in diameter with electrodes spaced at 8 mm,
- Drillability discharge category is classified for various rocks based on experimental data and references (12 categories altogether).

Since in different tests different impulse repetition frequencies were used, the drilling speed for all DR was reduced to a frequency of 10 impulses/s. This approach is correct because the drilling speed linearly increases with frequency in the range $f =$ 1–20 impulses/s. The depth of well drilling in the field conditions reached 47.5 m for the examined rocks. The depth of continuous well drilling was in the range 0.4–30 m. The well diameter was $D_{well} =$ 100–400 mm. The ambient temperature changed in the range from +35 to–50 °C.

Table 7.1 Classification of ED drilling rigs

Serial number	Drilling rig type	Drill head type	Characteristics of rock			Working liquid	Drilling conditions	Drilling speed, m/h	Speed of mechanic drilling (diamond, rotor), m/h
			Rock	Rock hardness ratio (f_r)	Drillability category				
1	DR-44	DH-50-8 DH-50-14	Marble, sandstone	4–6	V–VI	* Cable + transformer oil	Lab.	6–9	–
2	DR-57	DH-88-20	Micro-quartzite	16–18	X–XI	*Cable + transformer oil	Lab.	3.8–4.6	–
3	DR-73	DH-88-20	Micro-quartzite	16–18	X–XI	*Cable + transformer oil	Field	7.5	1.46
4	DR-73M	DH-120-20	Micro-quartzite	16–18	X–XI	**Oil	Field	3.1	1.36–2.0
5	DR-89	DH-110-15 DH-110-20	Micro-quartzite	16–18	X–XI	Diesel fuel	Field	2.2 2.9	–
6	DR-108	DH-140-20 DH-140-25 DH-140-30	Silicified sandstone	9–10	VIII–X	Diesel fuel	Field	1.3 2.1 5.6	–
7	DR-108	DH-150-25 DH-150-40	Micro-quartzite	16–18	X–XI	Diesel fuel	Field	3.0 4.2–6.7	–
8	DR-127	DH-200-30 DH-200-50 DH-200-75	Permafrost	7–8	VII–IX	Diesel fuel	Field	6.1 12.2 17.4	1.15–3.5
9	DR-426	DH-600-120	Silicified sandstone	9–10	VIII–X	Diesel fuel	Field	3.6	–

The first three types of drilling rigs (items 1–3 in Table 7.1) had not been developed due to extremely low reliability of their operation. The greatest volume of drilling was performed using drilling rigs 5–7, Table 7.1.

The DR-127 drilling rig (item 8, Table 7.1) was well designed based on previous long-term research and demonstrated very good results on the customer test site "Severovostokzoloto" that met the customer requirements. The customer was ready to invest in the development of a commercial drilling rig. However, the high school division (HVRI) alone could not solve this problem without cooperation with a powerful design organization and industrial enterprises.

No systematic investigation of drilling regimes with application of the DR-426 drilling rig (item 9 in Table 7.1) were performed, since after its testing in drilling wells for bridge supports, it was not in demand. The customer was not satisfied with the operating model because of strict time schedule for BAM construction. The customer was ready to pay big money (governmental!) for Japanese drilling rigs with diamond heads for mechanical drilling.

Table 7.1 presents drilling speeds of intermediate drillability rocks (with drillability category V–IX) using diamond and roller-bit drill heads by traditional mechanical methods. It can be seen that ED drilling of strong rocks even for small electrode spacing ($S = 20$ mm) provides a higher drilling speed. Along with other advantages (absence of rotation, greater drilling depth per head, etc.), it appears attractive for drilling deep wells of large diameter. In the test site near Tomsk, the drill head with electrodes spaced at 120 mm drilled the well 600 mm in diameter and a depth of 4.6 m in silicified sandstone. The average drilling speed was 3.6 m/h. This result confirmed the prospects for research and development of rigs intended for drilling wells of large diameter.

From Table 7.1 it follows that the main results of ED drilling were obtained using drill heads with electrodes spaced at 15–40 mm, which seriously limited transition to drilling of wells of large diameter. Thereupon, experiments on destruction of rocks by two-electrode rod system with decimeter electrode spacing were performed in the test site near Tomsk.

An analysis of the state of the art of drilling works in the world and of the stored experience of experimental EDT drilling indicate the necessity of EDT orientation toward drilling of wells with large diameter and mine wells. The development of ED systems for drilling mine wells 2000 mm in diameter using calc-bitumen solution as a flushing-out and working liquid gave the first encouraging result. The HVRI carried out these works in Donbas (because of disintegration of the USSR, the system was not brought to operating condition and remained in the territory of Ukraine).

An analysis of different modern methods of drilling deep wells 500 mm in diameter aimed at reduction of the cost of building geothermal wells led to the conclusion that the ED method of drillings has so far remained preferable despite of lacking of commercial systems implementing this method and progress in mechanical methods of drilling. It should be noted that no systematic studies of drilling regimes for large electrode spacing (more than 40 mm, Table 7.1) were performed in the period of the most intensive works on the development of the ED technology. Similar studies began in 1994 for drill heads 120–360 cm in diameter with electrodes spaced at 45–120 mm.

7.2 Technological Scheme of the Drilling Rig

In the process of the development of the ED drilling method, mobile (Fig. 7.2a) and stationary (Fig. 7.2b) drilling rigs that did not differ in principle from one another and consisted of the same parts shown in Fig. 7.2a have been created.

In drilling under field conditions, the impulse voltage generators (IVGs) with two–half-period charging scheme were used as ***sources of impulse voltage***. For long-tem operation with increased impulse repetition frequencies, the IVG was modernized—charging resistances were replaced by inductive elements. Generators demonstrated sufficiently reliable operation in drilling at frequencies up to 10 imp/s (problems of IVG application in EDT have been listed above). IVGs with different parameters were used: operating voltage $U_{oper} = 200–500$ kV, stored energy $W = 150–5500$ J, energy deposition $M = 4–61$ J/mm, current oscillation period $T = 5.5–20.0$ µs, time before breakdown $t_{br} = 0.25–0.5$ µs and impulse repetition frequency $f = 1–15$ impulses/s.

In the system whose scheme is shown in Fig. 7.2b, a high-voltage impulse from the IVG was transmitted through a cable (KPV-300) to a high-voltage trolley and through a sliding clutch and a flexible wire to a high-voltage input of the drilling rig. When the well was deepened, the sliding clutch moved together with the high-voltage input. The length of the high-voltage trolley corresponded to the length of the drilling rig section.

The ***drilling rig*** for ED drilling comprises a high-voltage input, a sectioned column of drilling pipelines, and a drill head and serves for transmission of high-voltage impulse to the well bottom and pumping of the flushing-out liquid [1, 2].

The high-voltage input represents polyethylene oil-filled insulator inside which the current wire is fixed. *The drilling column* represents a coaxial design. The external (grounded) pipes are casing pipes of prospecting assortment 108 mm in diameter, and the internal (current carrying) pipes are water pipes 3/4″ in diameter. The column is sectioned; the length of sections is 4 m with pin-and-socket connection of external pipe sections and plug socket connection of the current wire. The current wire is fixed inside the drilling column and is centered on its axis by means of bushings. Such column design allows up & down operations to be performed sufficiently simply and the length of the drilling rig to be increased. The weight of a 1 m drilling column makes 13 kg.

- *Drill head* (of radially-tangential type) whose photograph is shown in Fig. 7.1 comprises high-voltage and grounded systems of electrodes. The high-voltage system is fixed with the help of the head insulator that can move relative to the grounded system in the axial direction. This allows constant contact of the drill head with the well bottom to be provided. In the given (described) system, the head diameter was 140 mm, and the electrodes were spaced at 25–30 mm.

The ***flushing*** (circulating) ***system*** performs two functions, namely, it insulates the current wire from the grounded pipe of the rig and allows destroyed rock (drill cuttings) to be carried out at the surface. It includes a flushing pump, a pumping

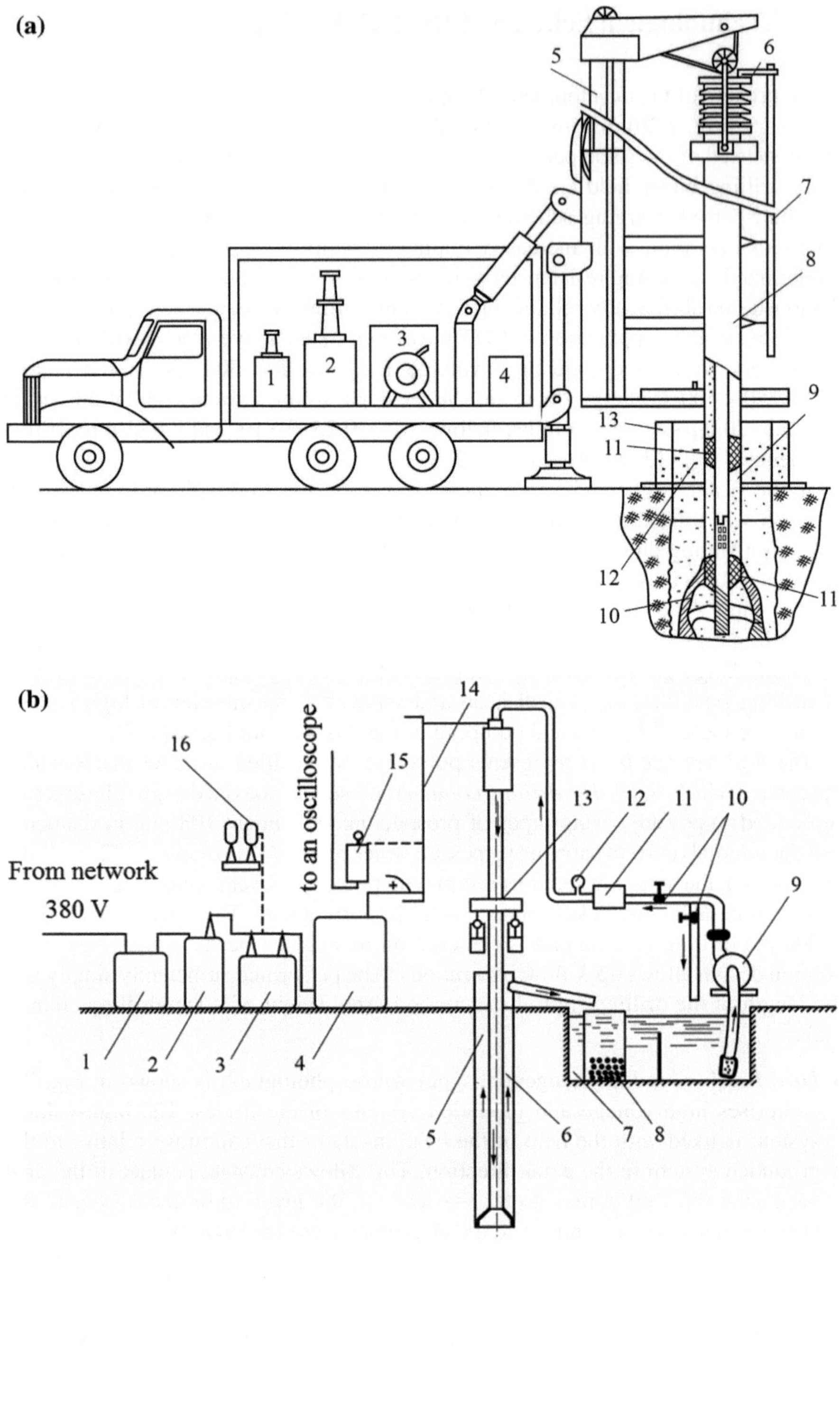
(a)
5
6
7
8
1
2
3
4
9
13
11
12
11
10
(b)
14
16
15
to an oscilloscope
From network
380 V
13
12
11
10
9
1
2
3
4
5
6
7
8

◂**Fig. 7.2** **a** Technological scheme of mobile ED drilling rig with reverse circulation of the flushing-out liquid by pumping comprising high-voltage charger 1, IVG 2, flushing device 3, portable control panel 4, mast 5, high-voltage input 6, high-voltage current collector 7, ED drilling rig 8, high-voltage electrical wire 9, drill head 10, insulators 11, and preventer 12. **b** Technological scheme of the stationary system for ED drilling comprising regulating transformer 1, transformer-rectifier bridge 2, throttle 3, IVG 4, drilling column 5, drilling column 6, sump 7, drill cutting tank 8, flushing pump 9, valve 10, flushing liquid flowmeter 11, manometer 12, device for pulling rig 13, high-voltage trolley 14, voltage divider 15, and kilovoltmeter 16

line, an annular space, a drain pipe with system for liquid removal from the well, a tank for flushing-out liquid, a system for emergency discharge of the flushing-out liquid, and a device for collecting drill cuttings and their subsequent removal from the circulating system.

In other systems, the direct scheme of flushing known from the experience of mechanical drilling was used in which the flow of the working liquid inside of the drilling rig was pumped with a pump through centering and head insulators and central high-voltage current wire, and the liquid backflow with destruction products flew in the space bounded by the external surface of the drilling rig and the wall of the well. This flushing scheme simple for implementation is often used in drilling by traditional methods. A disadvantage of this scheme of ED drilling is a large consumption of the flushing-out liquid. If hydrocarbons are used as flushing-out liquids, this is inadmissible by ecological and economic reasons. If the flushing-out liquid is water, the problem of using pumps with increased productivity and, as a consequence, increased power consumption becomes the most important.

- The ***up-and-down system of the drilling rig*** in ED drilling installations does not practically differ from that applied in mechanical drilling of wells and includes a drilling rig, a pile driver, and up-and-down appliances (elevator, folding forks, swivel keys, and clumps).

The ***control and measuring complex*** consisted of a low-inductive high-voltage divider of impulse voltage, a current shunt or Rogowski belt, a system of radio-frequency cables, a high-frequency impulse oscilloscope, and a frequency meter placed in a screened box. In addition, high rectified charging voltage of the IVG was monitored using electrostatic kilovoltmeters. The control system allowed continuous or periodic observations of breakdown and destruction processes in drilling to be performed. (Such complete control was carried out only in drilling in the testing site in the region of Tomsk.)

The ***control panel*** allowed all switching and adjustment operations in the course of drilling to be performed. In addition, devices protecting from the influence of high-voltage electromagnetic impulses—dividing transformers and rejector throttles—were used for components of the drilling rig that must be protected.

Exploitation of the ED drilling system belongs to works with increased level of danger since the personnel should deal with high voltage, combustible liquids, and load-lifting devices. The work must be done by a team of no less than three persons with observance of all rules of technical exploitation, safety precautions, and fire-prevention safety.

7.3 Operational Characteristics of Drill Rigs

7.3.1 Geometrical Parameters of the Drill Head

Design features of operating electrodes/drill heads determine in many respects the productivity and the energy consumption of rock destruction at the well bottom. Serious difficulties in the development of the drill head are caused by the necessity of optimization of its key parameters: sizes and shapes of electrodes, electrode spacing, mutual arrangement of electrodes in the drill head, and areas of end faces of drill heads. Such optimization required experimental investigations of ED drilling regimes for impulse energy and electrode spacing changing in wide limits.

To choose rational electrode sizes, two tendencies are observed: an increase in the area of the end face electrode surface is accompanied by the increase of their mechanical strength and lifetime of the drill head; on the other hand, this can lead to the formation of zones under electrodes with decreased breakdown probability and destructibility, because of which the drilling rig hangs.

As a rule, electrodes of two shapes—rod and plane ones—are considered when developing heads for ED drilling. To drill wells with large diameters, the original combination of rod and plate electrodes having different shapes turns out to be efficient, Fig. 5.2.

The drilling process is cyclic in character—descend of the drilling rig occurs discretely, that is, its motion alternates with pauses. The duration of pauses depends on the amplitude of voltage impulses and design of the drill heads. Complete termination of the drill rig descent can be caused by insufficient amplitude of voltage impulses for the given geometry of the drill head, local increase in the electric strength of rocks, or increase in the length of the discharge channel. Results of drilling with heads having electrodes of different shapes demonstrate that at $S = 20$ mm, the speed of drilling with plate electrodes was by 10–20% higher and the energy consumption by 15–45% lower than with heads having rod electrodes. For both electrode shapes and preset S and n values, an increase in the drill head diameter causes the increase of sizes (area) of the electrodes in contact with the well bottom and, as a consequence, the increase of pauses or termination of penetration ("hanging" of the drilling rig).

It was necessary to determine the influence of the electrode shapes on the destruction characteristics for heads with much larger electrode spacing ($S = 100$ mm) char-

Table 7.2 Granite destruction: $S = 100$ mm, indicated electrode systems, and average values of the measurable parameters

Electrodes	n, impulses	U_{br}, kV	I_{br}, kA	t_{br}, μs	W, J	W_s, J/mm	Q, cm^3/imp	W_s, J/cm^3	Sizes of slabbing crater			
									L, mm	B, mm	H, mm	V, cm^3
Rod–rod	1	601	16.2	0.44	5925	59.2	2.9	2049	38.3	20.7	3.8	2.7
	3	525	17.5	0.29	5978	59.8	16.7	357.6	110.4	73.4	9.3	50.5
	6	513	16.9	0.26	5892	58.9	18.7	315.0	130.4	96.5	35.8	112.8
	10	515	17.1	0.25	6114	61.1	17.5	349.4	147.0	107.0	41.0	175.0
Rod–plate	1	561	18.8	0.28	7202	72.0	13.4	537.5	93.0	52.0	14.4	13.4
	3	540	18.9	0.28	7115	71.1	28.9	246.2	145.0	61.6	21.3	86.7
	6	547	17.7	0.29	7026	70.2	26.6	264.1	159.2	98.4	36.0	160.0
	10	552	17.3	0.30	6861	68.6	29.7	231.0	158.3	146	40.2	297.5
	16	537	17.6	0.31	6826	68.2	23.3	292.7	178.0	176	48.0	372.0
Plate–plate	1	554	18.4	0.31	6219	62.2	2.6	249.7	43.0	14.3	6.0	2.6
	3	522	18.6	0.29	6145	61.1	33.4	193.6	133.6	94.6	21.6	100.3
	6	500	18.3	0.28	6268	62.7	33.5	195.0	146.6	165.0	31.3	201.6
	10	513	18.2	0.29	6297	63.0	40.0	162.9	183.3	207.6	40.6	400.0
	16	519	17.5	0.28	6138	61.4	33.4	183.8	180.0	210.0	46.0	550.0

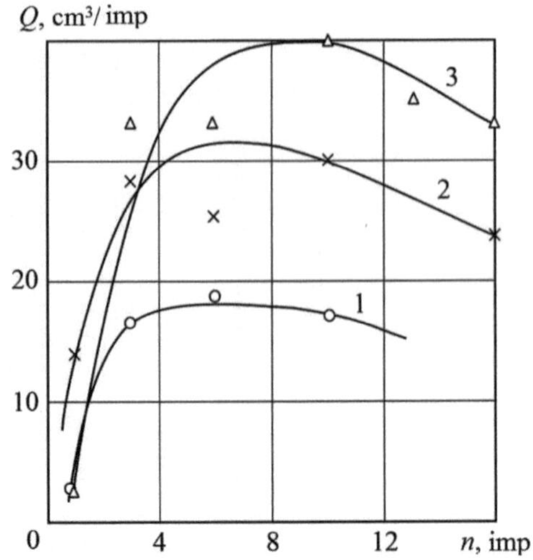

Fig. 7.3 Dependence of the destruction productivity on the number of impulses for electrode systems "core-core" (curve 1), "core-plate" (curve 2), and "plate-plate" (curve 3)

acteristic for drilling crowns of wells with large diameter. Three types of electrode systems were investigated: "core-core" with a core diameter of 10 mm, "core-plate" (core with a diameter of 10 mm and plate 250 mm long, 20 mm wide, and 5 mm thick), and "plate-plate" (with plates 90 mm long, 20 mm wide, and 5 mm thick). The core diameters and the widths and lengths of the plates are the operating parameters of the electrodes. For each electrode system, 3–7 experiments were performed. Results of experiments are generalized in Table 7.2 in which the average values of all examined destruction parameters and all electrode systems are given.

From Table 7.2 it can be seen that under repeated action of impulses, the breakdown voltages and the energy deposition for all electrode systems differ only slightly. The dependence of the productivity on the number of applied impulses is displayed by the curve with maximum for 6–10 impulses, Fig. 7.3. (The anomalous result of the action of the first impulse is explained by the measurement error, since no destruction can often be seen).

However, the most important thing is that for large electrode spacing the transition from the rod electrodes to the plane electrodes considerably increases the destruction productivity: for the "core-plate" electrode systems by a factor of 1.7 and for the "plate-plate" electrode system by a factor of 2.28. The average values of the relative destruction depths for all examined electrode systems and $n = 3$ impulses make $0.19S$–$0.23S$, Table 7.2. An increase in the number of impulses leads to the increase of the destruction depth. For $n = 16$ impulses, the average relative destruction depth reached $0.48S$, which is practically twice as much as for $n = 3$ impulses.

For insignificant difference between the destruction depths for different electrode systems, the destruction productivity significantly increases when going from the "core-core" electrodes to the "plate-plate" ones, Fig. 7.3.

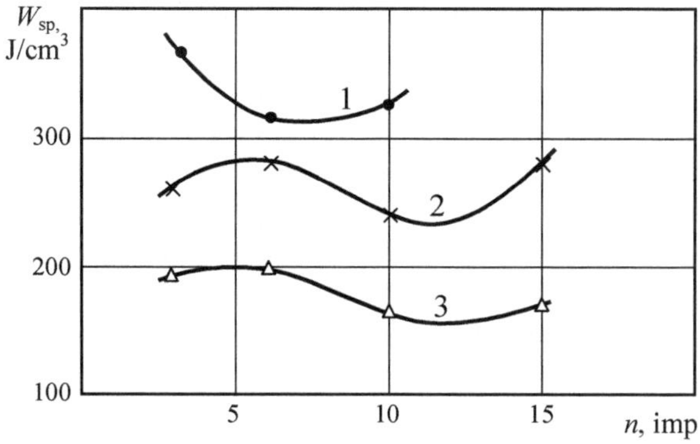

Fig. 7.4 Dependence of the energy consumption on the number of impulses for electrode systems "core-core" (curve 1), "core-plate" (curve 2), and "plate-plate" (curve 3)

This effect is primarily due to the increased destruction area (the width B of the slabbing crater) for the "plate-plate" electrodes and n = const. The difference increases with the number of applied impulses (Table 7.2).

The energy consumption—one of the main characteristics of the destruction process—is shown in Fig. 7.4 and given in Table 7.2. The considerable decrease of the energy consumption (by a factor of 2.14) is observed when going from the "core-core" electrodes to "plate-plate" ones.

Based on the results obtained, we can draw the following conclusions:

(1) The electrode shape influences the destruction characteristics of rocks,
(2) The maximal destruction productivity is observed for 6–10 impulses for the examined electrode shapes,
(3) The plate electrodes provide much better destruction characteristics than the rod electrodes,
(4) From Table 7.2 it follows that the length of the slabbing craters is larger than the electrode spacing upon repeated exposure to voltage impulses. This allows us to assume that the electrode sizes (the core diameter and the plate width) can be larger than those used in the above-described experiments.

Based on the results of tests of crowns with different geometrical parameters, the following recommendation can be formulated: when designing the drilling crown, we should strive to the maximal area of the end face electrode sections, but they must not exceed admissible value preset by the condition of complete utilization of the sub-electrode volume, that is, prevent electrode hanging. To meet these requirements in crowns with rod electrodes, the diameter of the electrodes should be chosen from the condition $d_{electr} \cong 0.7S$, where S is the electrode spacing. The application of electrodes of different cross sections in a homogeneous head is undesirable because of inhomogeneous energy distribution in the spark channels over the well bottom. The destruction is more intensive near the electrode with a smaller cross section. These recommendations refer to the drill heads intended for drilling wells with relatively small diameter (100–150 mm) and, as a rule, for the cylindrical (rod) shape of the electrodes and their small spacing. For drilling deep wells (3–4 km) of large diameters (up to 400 mm), the crowns of larger diameter with greater spacing and hence, with electrodes of different shapes and sizes are required.

To design such heads, later on (from the late 80s) the influence of their geometry on the rock destruction characteristics was investigated. (The most part of field tests were performed under the leadership of N. E. Kovalenko, and laboratory investigations were performed under the leadership of V. F. Vazhov.) The slabbing crater parameters for core electrodes 8–20 mm in diameter or for plates with sizes of 60 × 40 × 15 mm superimposed on one sample surface (Fig. 7.5) were measured.

The length of the slabbing crater under the electrode was calculated from the expression $L_1 = L_2 = (L - S)/2$ [mm], where L_1 and L_2 are the lengths of slabbing craters under the electrodes, in mm; L is the maximum length of the slabbing craters, and S is the electrode spacing. The relative length of the slabbing crater under the electrode $K = L_1/S$ was also calculated.

The time before breakdown in all cases was in the range 0.3–0.5 μs. Many experiments on granite destruction were performed with rod electrodes having the diameter $d_{electr} = 8$ mm, during which the electrode spacing ($S = 40$–120 mm), the operating voltage ($U_{oper} = 306$–400 kV), the discharge energy ($W_{disch} = 320$–1038 J), and the number of impulses ($n = 3$–12 impulses) were varied.

For the rod electrodes 8 mm in diameter, the absolute length of the slabbing crater under the electrode L_1 was practically independent of the electrode spacing (line 1 in Fig. 7.6), and its relative value L_1/S decreased with increasing S (curve 2 in Fig. 7.6). The length of the slabbing crater under the electrode in this case was determined by the electrode diameter, since the average absolute L_1 value exceeded the electrode diameter by 6%, which was caused by the stochastic character of the process of forming the slabbing crater in the rock massive and did not exceeded the measurement accuracy.

These results have demonstrated that for larger crowns, the electrode diameter must be chosen from the condition $d_{electr} \leq 0.7S$. This relationship was derived for $S = 20$ mm, that is, for $d_{electr} \leq 14$ mm. Approximation of curves 1 and 2 in Fig. 7.6 for $S = 20$ mm yields $K = L_1/S = 0.4$, and setting $d_{electr} = 14$ mm, we obtain $K = 0.7$.

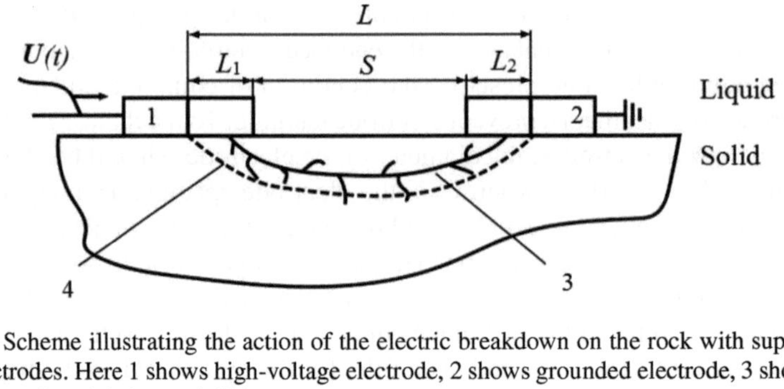

Fig. 7.5 Scheme illustrating the action of the electric breakdown on the rock with superimposed plate electrodes. Here 1 shows high-voltage electrode, 2 shows grounded electrode, 3 shows breakdown channel, and 4 shows slabbing crater

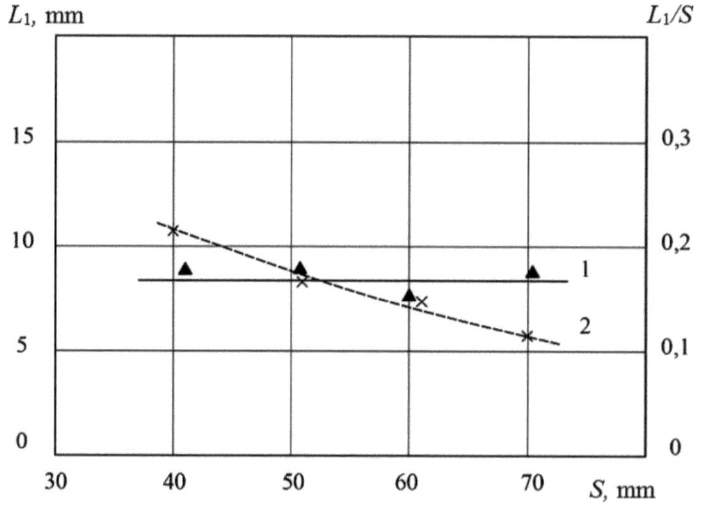

Fig. 7.6 Dependence of the length of the slabbing crater under the electrode on the electrode spacing for granite. Here line 1 is for L_1 and curve 2 is for L_1/S

It was also established that it is expedient to use the drill heads with rod electrodes to drill wells with diameters no more than 200 mm; for larger well diameters, the plate electrodes are more efficient. This conclusion is based on the experiments performed with samples made of chamotte brick (the mechanical strength coefficient $f = 4$–6) and changing operating voltage and electrode spacing. For all operating voltages ($U_{oper} = 90$–190 kV), the relative and absolute lengths of slabbing craters fast decreased with increasing electrode spacing at U_{oper} = const, Fig. 7.7. For constant S, it increased with U_{oper}.

For example, for $S = 20$ mm, the length of the slabbing crater under the electrode increased with operating voltage in the examined range (curve 5 in Fig. 7.8). This result was confirmed by experiments with samples made of ferriferous quartzite ($f = 12$–14), curve 7 in Fig. 7.8, at $U_{oper} \leq 200$ kV, $K > 2$. Hence it follows that for larger crowns, the end face area of the electrodes at the bottom of the well can also be greater than for crowns of smaller diameter.

Drilling of sandstone ($f = 6$–8) with an actual drill head having plate electrodes demonstrated that for $S = 14$ mm and $U_{oper} = 280$ kV, the relative length of the slabbing crater under the electrode was $K = 1.86$.

From Fig. 7.8 it can be seen that with increasing electrode sizes, the increase of the operating voltage causes the corresponding increase of the length of the slabbing crater under the electrode L_1 for all interelectrode distances. For example, doubling of the operating voltage (from 100 to 200 kV) increased the length of the slabbing crater by a factor of 2.9 for $S = 2.5$ mm, and for $S = 25$ mm, it increased by a factor of 4.3, that is, the effect of increase in the length of the slabbing crater with increasing electrode spacing is observed. In addition, the tendency toward the

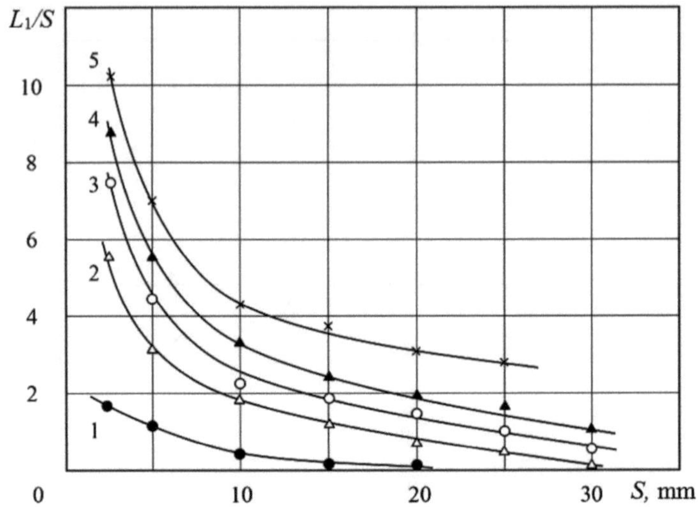

Fig. 7.7 Dependence of the relative length of the slabbing crater under the electrode on the electrode spacing for chamotte brick and voltage impulse amplitudes (U_{oper}) of 90 (curve 1), 110 (curve 2), 130 (curve 3), 150 (curve 4), and 190 kV (curve 5)

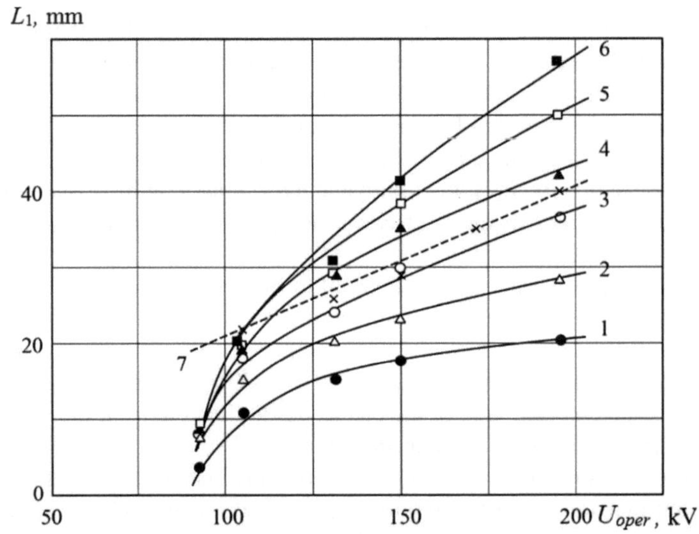

Fig. 7.8 Dependence of the length of the slabbing crater under the electrode on the voltage for chamotte brick (curves 1–6) and ferriferous quartzite (curve 7) at electrode spacing of 2.5 (curve 1), 5 (curve 2), 10 (curve 3), 15 (curve 4), 20 (curve 5), 25 (curve 6), and 20 mm (curve 7)

increase of the length of the slabbing crater under the electrode was observed when the operating voltage increased and then saturated. Hence, the length of the slabbing crater and hence, the electrode area are limited. This effect can be explained by the decreased breakdown probability with increasing length of the discharge in a solid. The electrode size determining the effective crown operation should be chosen by a comparison of the voltage-time characteristics (VTC) of the solid being destructed and the working liquid, namely, the greater the difference between $U_{br.sol}$ and $U_{br.l}$, the larger can be the electrode size.

Several modifications of drilling rigs (the scheme of one of them is shown in Fig. 7.9) and drill heads intended for drilling of wells with large diameters were developed, Fig. 7.10.

Their main constructive sizes are given in Table 7.3 by analogy with Table 7.1. The optimal ratio of the diameters of coaxial electrode system from the viewpoint of one- and multi-impulse electric insulation strength in the drilling rig is in the range 2.78–3.3, that is, d_{in}/d_{out} values in Table 7.3 lie practically in this range.

The problem of reliable operation of insulation of the drilling rig is very urgent due to the high electric field strength. The solid insulation operates under extreme conditions taking into account that the turbulent flow of the polluted working liquid flows through the insulators. A great volume of calculations and experiments was required to design high-voltage inputs and insulators of the drill head intended for drilling rigs that meet the EDT requirements, Table 7.3. In drilling, breakdowns and flashovers over the insulator surface were observed for the DH-120-35 and DH-120-45 heads of the BI-89 drilling rig due to very high working gradients: $E_{oper} =$ 21.7 kV/mm. When drilling with DH-270-70 and DH-360-120 rigs, no insulation failures were observed, since the field strength in them did not exceed 11.7 kV/mm.

Only a portion of energy stored in IVG is released in the discharge channel. The corresponding discharge circuit efficiency η depends on the relative contribution of the active component of the discharge channel resistance to the total resistance of the discharge circuit that increases with the discharge gap length S. The length of the discharge channel and its active resistance change from discharge to discharge for preset electrode spacing S in a random way, which forces us to use a certain average calculated value of the discharge channel resistance in actual field regime of drilling. For discharge gap lengths in the range 10–120 mm, $\eta = 0.65$–0.75 [3]. The portion of energy liberated in the discharge channel at $S = 20$ mm in microquartzite, granite, and sandstone makes $\eta = 0.37$–0.56, increasing from sandstone to microquartzite. In drilling under field conditions, it is convenient to estimate the energy parameters from the energy stored in the IVG.

The energy released in the discharge channel was determined from the waveforms of the current and voltage impulses for the first half-cycle of the discharge current. To this end, a computer program was used. Figure 7.11 shows typical voltage and current waveforms in drilling. Based on numerous calculations of the energy stored in the discharge channel (thousands of measurements), we can assume that the transmission coefficient η of energy stored in IVG to the discharge channel is the following: $\eta =$ 0.52–0.67 for $S = 80$ mm and $\eta = 0.59$–0.73 for $S = 120$ mm.

There exist nearly 40 criteria for the drilling efficiency [4, 5], but there is no universal criterion. Each criterion has its own application field, characterizes the technical mean from a certain side, and must be used in combination with others. To estimate the ED drilling, the following criteria were chosen:

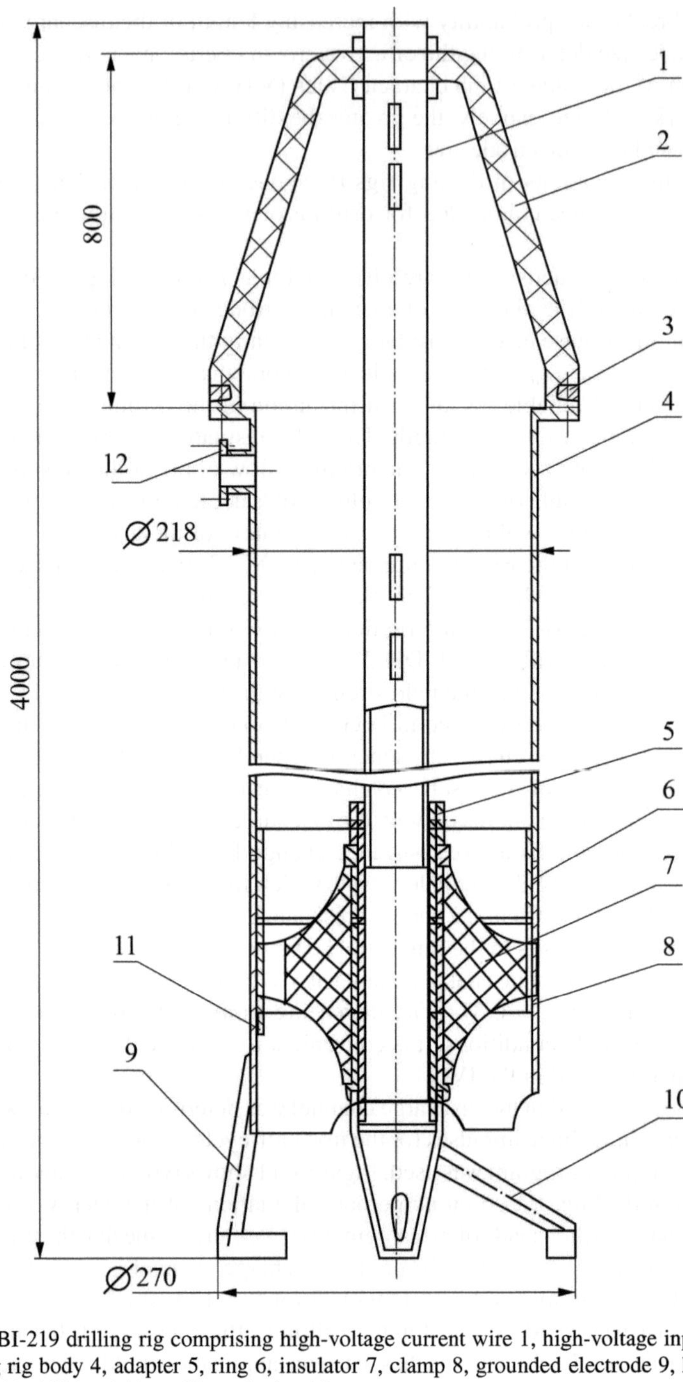

Fig. 7.9 BI-219 drilling rig comprising high-voltage current wire 1, high-voltage input 2, flange 3, drilling rig body 4, adapter 5, ring 6, insulator 7, clamp 8, grounded electrode 9, high-voltage electrode 10, drill head 11, and branch pipe 12

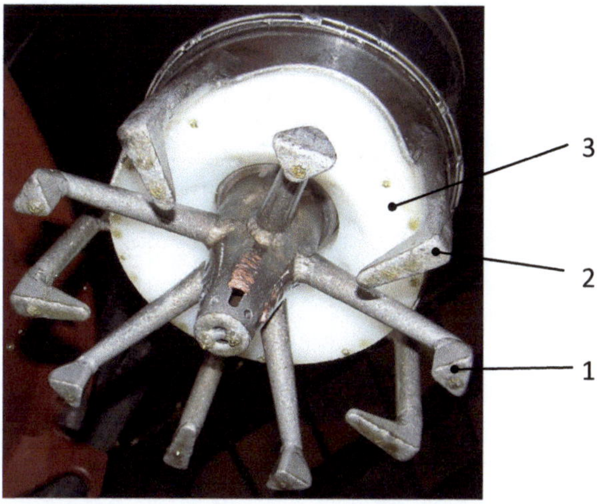

Fig. 7.10 DH-270-70 drill head 270 mm in diameter comprising high-voltage electrode 1, grounded electrode 2, and insulator 3

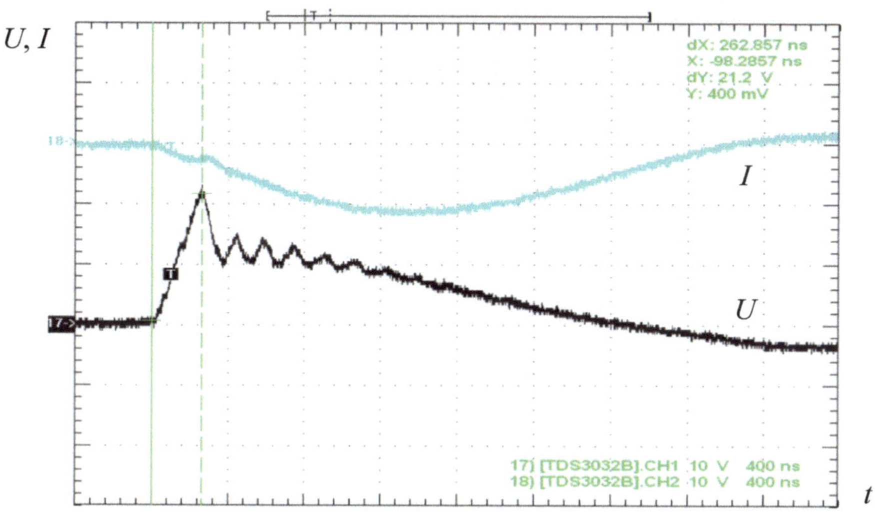

Fig. 7.11 Waveforms of voltage and current impulses in drilling with a BI-325 drill rig

- The breakdown voltage U_{br} determined from the waveforms,
- Drilling speed V,
- Drilling productivity $Q = \frac{P}{n}$, where P is the volume of the drilled well and n is the number of acting impulses,
- Specific energy consumption $W_{spec} = \frac{W_{\Sigma}}{P}$, where $W_{\sum} = W{\cdot}n$,
  Energy deposition $W_s = W/S$.

Table 7.3 Characteristics of drilling rigs

Drilling rig type	Drill head type	d_{in}/d_{out}	Characteristic of rock			Working liquid	Breakdown voltage, kV	V_{10}, m/h
			Rock	Strength coeff. f_{str}	Drillability category			
BI-89	DH-120-35 DH-120-45	80/27	Granite	14–16	IX–X	Transformer oil	275/316	5.4 6.0
BI-219	DH-270-70	209/60	Granite	14–16	IX–X	Transformer oil	437	7.6
BI-325	DH-360-80 DH-360-120	310/108	Granite	14–16	IX–X	Transformer oil	436/492	3.7 4.8

Note Here d_{in} is the internal diameter of the drilling pipe column, and d_{out} is the external diameter of the high-voltage current wire

7.3.2 Operating Voltage

The operating voltage amplitude for ED drilling depends on the electrode spacing, properties of a solid and liquid, as well as on some other parameters. Nowadays it is impossible to calculate it analytically, especially taking into account the special features of the rock. Therefore, experimental results and empirical dependences provide the basis here. High degree of inhomogeneity of rocks causes high voltage spread. In addition, the length of the discharge channel is greater than the electrode spacing, because the channel penetrates into a solid to depths that are stochastic in nature.

However, the results accumulated by the present time allow one to judge about the amplitude of voltage necessary for drilling in different conditions. Figure 7.12 shows values of the breakdown voltage depending on the electrode spacing S for drill heads of different designs and diameters and some rocks—microquartzites, granites, sandstones, and permafrost. Attention is drawn to their wide spread even for the same rock type.

The results shown by curve 2 in Fig. 7.12 demonstrate the weak dependence of the average breakdown voltage for different rocks caused by the strong influence of real drilling conditions on the electric strength, first of all, of multi-impulse action on the rock causing intensive cracking at the well bottom and walls and high rock surface roughness. The dependence of the breakdown voltage on the electrode spacing reflects the well-known "effect of total voltage," according to which U_{br} nonlinearly (more slowly) increases with S: the Seven-fold increase in S leads to the increase of the maximal U_{br} value by a factor of 2.16 that allows heads with large electrode spacing to be used. For more effective destruction (with increased probability and penetration depth of the discharge channel into the rock), the maximal U_{br} values must be used, see curve 1 in Fig. 7.12 calculated by the empirical expression

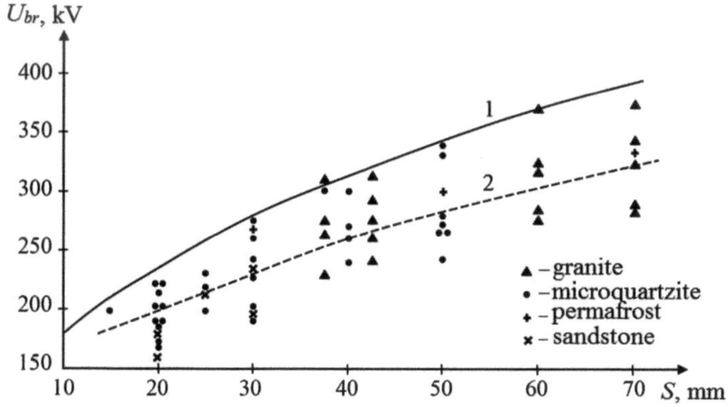

Fig. 7.12 Dependence of the breakdown voltage on the electrode spacing for indicated rocks and well diameters. Here curve 1 shows maximal U_{br} values (calculated from Eq. (7.1)), and curve 2 shows the average U_{br} values

$$U_{br} = U_{\max} \cdot S^{0.4}, \qquad (7.1)$$

where $U_{\max} = 180$ kV is the maximal breakdown voltage at $S = 1.0$ cm, 0.4 is the experimental coefficient, and S is the electrode spacing. The exponent in the dependence $U_{br} = f(s)$ for different dielectrics and rocks changes in the range 0.33–0.6.

Drilling of granite with a BI-325 rig equipped with DH-360-80 and DH-360-120 drill heads demonstrated that the breakdown voltage (Table 7.2) is satisfactory described by Eq. (7.1). Results of calculations by Eqs. (2.9) and (7.1) are both acceptable taking into account the spread of breakdown voltages. In this case, the possibility arises to carry out laboratory experiments with a two-electrode system in compliance with required methodical recommendations to refine U_{br} for concrete drilling conditions in case of need.

7.3.3 *Particle Size Distribution and Fractional Composition of Drill Cuttings*

To choose equipment for ED drilling of wells, it is necessary to have the detailed characteristic of rock cuttings as a whole and of its largest fragments that must be extracted from the well bottom zone to the surface.

The well penetration consists of two main stages—rock breaking from the massive and removal of the broken rock at the surface. Conditions of rock destruction, cleaning of the well bottom, and removal of the destructed rock at the surface in ED drilling differ significantly from analogous processes in traditional mechanical drilling.

The indispensable condition of effective ED drilling is timely and complete cleaning of near-bottom zones from products of rock destruction and decomposition of flushing-out liquids during time interval between impulses. Otherwise, secondary crushing (recrushing) of cuttings will be observed or the subsequent discharges will occur through gas-filled cavities. All this decreases the productivity, increases the energy consumption, and decreases the drilling speed.

To choose conditions of removal of drill cuttings (the flushing intensity), their size distribution and fractional composition must be known, since they largely determine the design of the drill head and drilling rig limiting the repetition frequency of impulses applied to the well bottom. It is necessary not only to eliminate recrushing of drill cuttings, but also jamming of the drilling rig in the well for direct flushing or clogging of the pipe in the well with reverse flushing.

At present it is impossible to determine *a priori* the expected characteristics of drill cuttings and the parameters of the flushing system of the well. Corresponding experimental research and generalization of the accumulated empirical material were required. The energy consumption was accepted as the energy characteristic of destruction when analyzing drilling cuttings in wells of large diameter with electrode

Table 7.4 Relative maximal sizes of drill cutting fragments

Rock	S, mm	W_s, J/mm	l_{max}/S	b_{max}/S	h_{max}/S
	Destruction with a two-electrode system				
Granite	60	10.2	1.15	0.65	0.27
		20.4	0.92	0.72	0.38
		30.0	1.03	0.67	0.25
		48.0	0.81	0.42	0.28
	90	20.6	1.08	0.70	0.41
	120	15.0	1.02	0.76	0.26
		20.1	1.01	0.63	0.27
	Well drilling with a drilling head				
Granite	60	5.2	1.17	0.95	0.32
		7.5	1.23	0.73	0.33
		10.2	1.25	0.95	0.38
		15.0	1.03	0.75	0.30
		18.1	0.95	0.73	0.32
		19.0	1.06	0.58	0.32
		20.4	0.97	0.77	0.30

spacing exceeding 45 mm. The main experiments were performed for well drilling in microquartzite, granite, sandstone, and limestone.

In the laboratory studies, the two-electrode system superimposed on one surface of the rock sample placed in transformer oil was used. The electrode diameter was 10 mm. The contact of the electrode surface with the sample surface was flat cut. The working medium was diesel oil (for drilling in field conditions) or transformer oil (for drilling in laboratory conditions).

ED drilling of wells was performed using drilling rigs with multi-electrode drill heads and different values of the electrode spacing S. The destructed rock (drill cuttings) was extracted from the well with diesel oil to the surface and put in a sump.

After termination of drilling, drill cuttings were extracted from the sump, dried, and sifted through standard sieves to subdivide into fractions. The diameter of the sieve mesh changed from 5 to 20 mm. The fractions were weighed, and percentage of each fraction P was determined. In the largest fraction, the largest fragments characterized by the average length $\bar{l}$, width $\bar{b}$, and thickness $\bar{h}$ were chosen. In this case, the result inevitably depended on the method of determining average sizes: the larger number of fragments was chosen in the given experiment, the smaller were the sizes obtained. For measurements, 10 largest fragments were chosen. One fragment with maximal sizes was chosen from them, and its length l_{max}, width b_{max}, and thickness h_{max} were measured. The measurement accuracy was ± 0.1 mm.

In the laboratory studies, the sizes of rock fragments were measured after exposure to each voltage impulse.

As indicated above, the special feature of ED drilling of rocks is that they break into large fragments. This is confirmed by observations of the relative maximal sizes of individual fragments of drilling cuttings as functions of the electrode spacing both for breaking by two electrodes and drilling wells for values of deposited energy W_s indicated in Table 7.4. It can be seen that the change of S and W_s does not practically influence on the relative sizes $l_{\max}$, $b_{\max}$, and $h_{\max}$. The maximal length of the fragment is primarily determined by the length of the discharge channel; for the indicated S values, the length of fragments $l_{\max} \leq 1.25S$ is expected. The width of the largest fragment is a random value determined primarily by the discharge channel trajectory and properties of the rock (its strength, brittleness, mineralogical structure, etc.). From the results obtained it follows that the expected maximal width is $b_{\max} \leq 0.95S$. The thickness of the fragment is determined by the depth of the discharge channel penetration into the rock. The most probable penetration depth makes ~$0.32S$ [6]. However, the thickness of the fragments was much greater ($h_{\max} \leq 0.4S$). Proceeding from the given values of the maximal sizes of drill cutting fragments, we can state that they have flaky structure and hence, the greatest sizes during flushing will be $b_{\max}$ and $h_{\max}$.

The occurrence of drill cutting fragments with maximal sizes is a random and relatively rare event. In this connection, it is interesting to analyze the average sizes of the largest drill cutting fragments excluding extremely large ones. Figure 7.13 shows the dependences of the average sizes of the largest drill cutting fragments on the electrode spacing for the energy deposition $W_s = 18$–22 J/mm for granite and microquartzite obtained in experiments with the two-electrode system and in drilling. It can be seen that in all cases, change of $\bar{l}$, $\bar{b}$, and $\bar{h} = f(S)$ for granite can be described by linear dependences of the form

$$\bar{l} = a \cdot S, \bar{b} = c \cdot S, \bar{h} = k \cdot S, \tag{7.2}$$

where a, c, and k are experimental coefficients depending on the rock properties. For granite $a = 0.8$, $c = 0.55$, and $k = 0.25$, and for microquartzite they are 0.7, 0.46, and 0.23, respectively. In Fig. 7.13, deviations of $\bar{l}$ and $\bar{b}$ from linear dependences caused by the methodical features of measurements can be seen. The satisfactory coincidence for $\bar{h}$ is caused by the weak dependence of the discharge channel penetration depth on physicomechanical properties of rocks.

We can state *a priori* that the change of the energy deposited into the discharge channel should influence on the size distribution of drill cuttings in ED destruction. As demonstrated above, the energy deposition does not influence on the maximal sizes of individual particles; however, it affects the average values of $\bar{l}, \bar{b},$ and $\bar{h}$. Fig. 7.13 shows the dependences of the average values of $\bar{l}, \bar{b},$ and $\bar{h}$ on the energy deposition into the discharge channel for granite at $S = 60$ mm in experiments with two electrodes and in drilling.

It can be seen that a fivefold increase in the energy deposition causes the decrease of the average sizes of $\bar{l}, \bar{b},$ and $\bar{h}$ by factors of 1.46, 1.8, and 1.6, respectively. Analogous dependences for granite in the examined range of W_s can be described by linear equations:

$$\bar{l} = \bar{l}_0 - a_1 \cdot W_s, \bar{b} = \bar{b}_0 - c_1 \cdot W_s, \bar{h} = \bar{h} - k_1 \cdot W_s, \tag{7.3}$$

where $\bar{l}_0 = 55$ mm, $\bar{b}_0 = 40$ mm, and $\bar{h}_0 = 18$ mm and $a_1 = 0.375$, $c_1 = 045$, and $k_1 = 0.15$ for granite at $S = 60$ mm.

From Fig. 7.14 and Eq. (7.3) it follows that changing the energy deposited into the discharge channel, the average sizes of the largest drill cutting fragments can be regulated.

From the viewpoint of optimization of well flushing, it is important to know the fractional composition of drill cuttings depending on the drilling conditions. Figure 7.15a shows the dependences of the relative weights of different fractions on the electrode spacing for microquartzite ($S < 30$ mm, the fraction +20 mm was absent), and Fig. 7.15b shows these dependences for granite.

In the course of solid destruction using the spark discharge, four successive stages are identified: (1) growth of a great number of radial cracks, (2) increase in the number of cracks with energy deposition rate, (3) sharp decrease of the number of growing cracks with rate of their growth remaining maximal, (4) several single growing cracks are retained that after imposition of the electrodes rise at the surface of the solid [3, 7]. The presence and volume of small fractions are primarily determined by the first two stages, and the size of the largest fragments of the rock being destructed is determined by the number of cracks left on the free surface. Their number correlates with the energy deposited into the channel.

It is well known that the size distribution of drill cuttings depends on the electrode spacing. For example, for $S \leq 30$ mm, no drill cuttings with sizes of 20 mm and

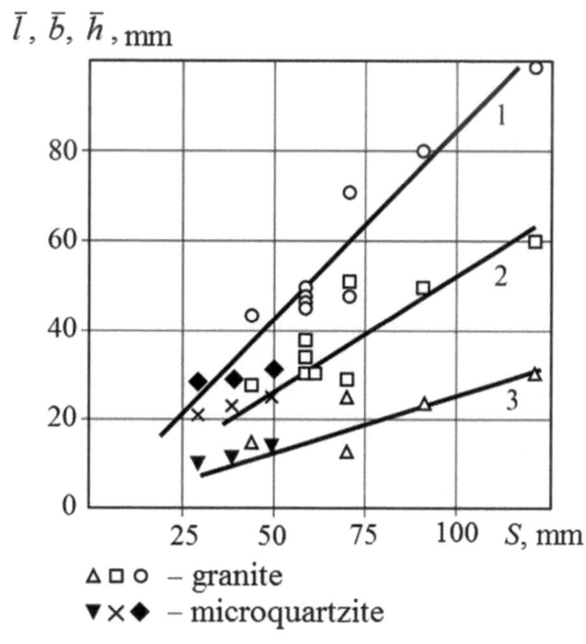

Fig. 7.13 Dependences of the average maximal sizes of drill cutting fragments on the electrode spacing for microquartzite and granite at W_s = 18–22 J/mm. Here curve 1 is for $\bar{l}$, curve 2 is for $\bar{b}$, and curve 3 is for $\bar{h}$

Fig. 7.14 Dependence of the average values of maximal sizes of drill cutting fragments on the energy deposition for granite with $S = 60$ mm. Here curve 1 is for $\bar{l}$, curve 2 is for $\bar{b}$, and curve 3 is for $\bar{h}$

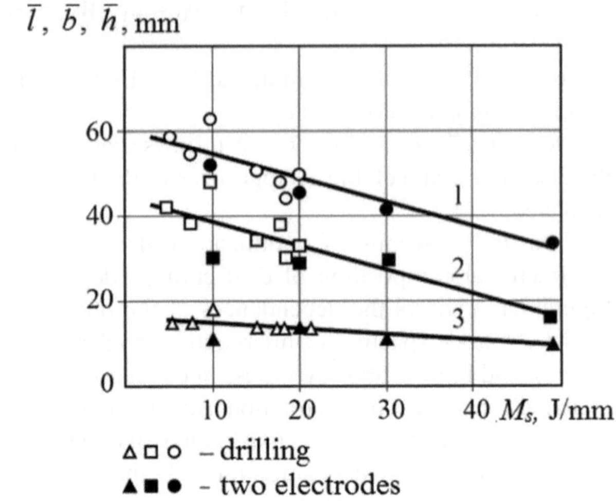

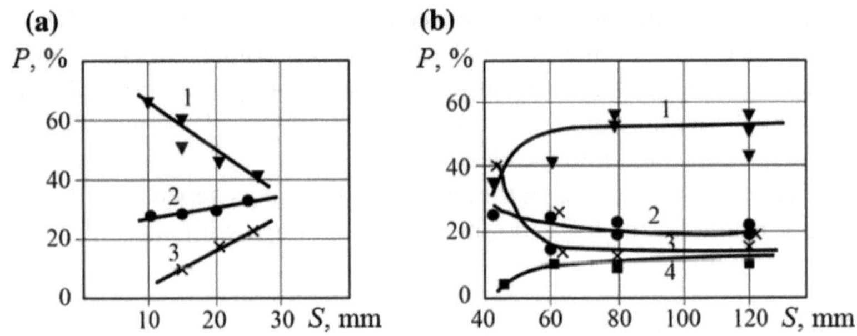

Fig. 7.15 Dependences of the fractional composition of drill cutting on the electrode spacing for W_s = 20–35 J/mm: **a** microquartzite drilling and **b** granite drilling. Here curves 1 are for fraction −5 mm, curve 2 is for fraction +5 mm, curve 3 is for fraction +10 mm, and curve 4 is for fraction +20 mm

larger were observed, and drill cuttings with sizes of 5 mm and larger could be considered large, whereas for $S > 45$ mm, drill cuttings with sizes exceeding 20 mm could be considered large. Below, in connection with the uncertainty of drill cutting classification into "large-small," a drill cutting fraction of −5 mm is considered small, and a fraction of +20 mm is considered large. Depending on S, fractions of intermediate sizes +5 mm and +10 mm can be considered as small (for $S > 45$ mm) or large (for $S < 30$ mm). Then for the examined range of electrode spacing, we can state that with increasing S, the relative fraction of small drill cuttings fast decreases (curve 1 in Fig. 7.15a), whereas the relative fraction of large drill cuttings increases (curves 2 and 3 in Fig. 7.15a).

Figure 7.15b shows the dependences of the relative values (in %) of fractions of drill cuttings in drilling wells of large diameter (D_{well} = 300–400 mm) for the indicated values of spacing of electrodes in the drill head (S = 45–120 mm) and energy deposition ($W_s = 20$–35 J/mm for $S = 45$–60 mm and $W_s = 40$–61 J/mm for $S = 80$–120 mm). For $S > 80$ mm, the relative contribution of very small drill cuttings (−5 mm) increased for which the first and second stages of recrushing and cracking are responsible for energy deposition into the discharge channel. This is most likely caused by a significant increase (by a factor of 1.74–2.0) of energy deposition for $S = 80$–120 mm, which inevitably causes an increase in the diameter of the region of material recrushing near the discharge channel [8]. This is confirmed by the data shown in Fig. 7.15. From the figure it can be seen that the relative contribution of other small fractions (+5 and +10 mm) for the examined electrode spacing decreases, whereas the relative contribution of large fractions (+20 mm) increases by a factor of 5.6 when S increases by a factor of 2.7, by analogy with microquartzite (see curve 3 in Fig. 7.15a). These dependences demonstrate that the energy efficiency of ED drilling significantly increases with S, especially for $S > 45$ mm, since the relative contribution of large fractions increases.

The average size of the fragment of the destructed material decreases with increasing energy. The dependence of the fractional composition of drill cuttings in drilling of microquartzite with $S = 20$ mm on the energy deposition is shown in Fig. 7.16.

It can be seen that the relative contribution of small fraction (–5 mm) increases, whereas the relative contribution of large fractions (+5 and +10 mm) decreases. As the energy deposition increases by a factor of 4.1, the relative contribution of the small fraction increases by 10%, and for large fractions, it decreases by 6%, that is, the increase of the energy deposition influences on the size distribution of drill cuttings.

Figure 7.17 shows the dependences of the fractional composition of drill cuttings on the hardness ratio for sandstone ($f_p = 8$), limestone ($f_p = 10$), granite ($f_p = 16$), and microquartzite ($f_p = 18$).

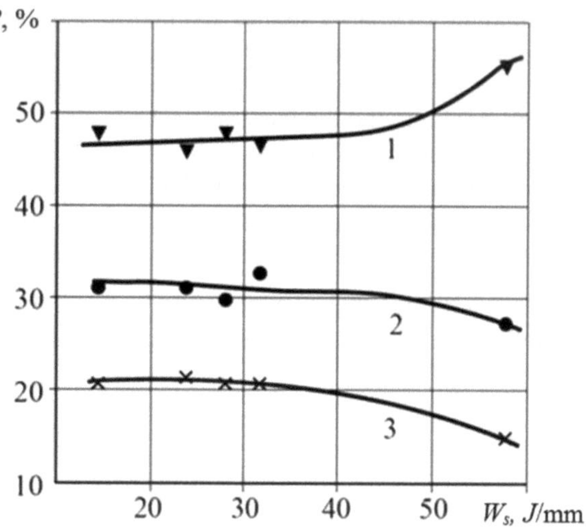

Fig. 7.16 Dependence of the fractional composition of drill cuttings on the amount of energy deposition in drilling of microquartzite with S = 20 mm. Here curve 1 shows a fraction of −5 mm, curve 2 shows a fraction of +5 mm, and curve 3 shows a fraction of +10 mm

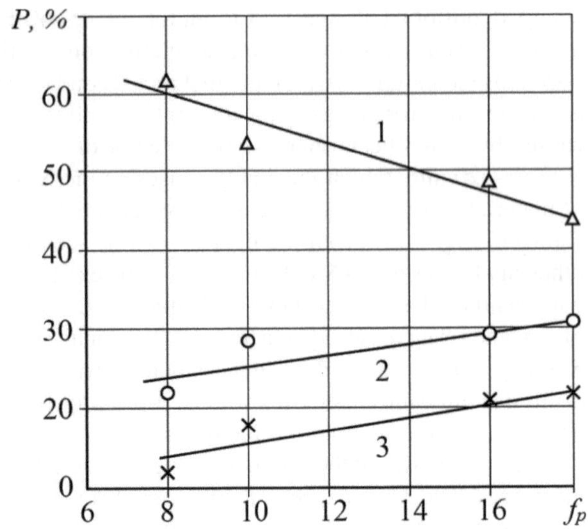

Fig. 7.17 Dependences of the fractional composition of drill cuttings in drilling on the rock-hardness ratio (f_p) for S = 20 mm and W_s = 20–50 J/mm. Here curve 1 is for a fraction of −5 mm, curve 2 is for a fraction of +5 mm, and curve 3 is for a fraction of +10 mm

The increase in rock-hardness ratio by a factor of 2.25 reduces the relative contribution of small fraction (−5 mm) by 19% and increases the relative contribution of large fractions (+5 mm and +10 mm) by 9 and 10%, respectively.

Summing up the investigations described above, we can draw the following conclusions:

- Maximal size of drill cutting fragments is directly proportional to the electrode spacing and is practically independent of the energy deposited into the discharge channel,
- Average sizes of the largest fragments decrease with increasing energy deposition both in the laboratory studies and in drilling of actual wells,
- The composition of drill cuttings in ED drilling depends primarily on the electrode spacing: an increase in spacing leads to the decrease of the relative contribution of small fractions and the increase of the relative contribution of the large fraction,
- For increased rock-hardness ratio, the smaller relative contribution of small fractions and the large relative contribution of the large fraction is characteristic,
- The increase of the energy deposition causes the relative contribution of the small fractions to increase and of the large fractions to decrease.

The results obtained allow the system of well flushing to be chosen and calculated as well as the elements of the drilling rig to be designed to increase the drilling efficiency.

7.3.4 Intensity of Well Flushing

Cleaning of the well from the destruction products is the indispensable condition of any drilling method and can be performed either with mechanical devices (screws or sludge pumps) or in a flow of a flushing agent (air flushing or purging). Nowadays, the well cleaning in ED drilling is carried out by flushing.

Unlike mechanical drilling, in ED drilling a high probability of forming large fragments of drill cuttings (exceeding 70 mm) and vapor-gas products of decomposition of flushing-out liquids at the well bottom are observed [9]. These special features of the EDT in well drilling influence significantly the pulp pipe parameters and the flushing regime.

The efficiency of flushing can be estimated using such criterion, as the energy consumption, that is, the energy consumed to obtain the required rate of flushing-out liquid flow through the pulp pipe and other elements of a circulating system, and the material consumption, that is, specific consumption caused by technological losses of the flushing-out liquid. The energy consumption for flushing decreases with power spent for its implementation and with increasing drilling speed. Hence, to optimize the energy consumed for flushing, the flushing parameters must be chosen so that with other parameters remaining the same, the flushing power was minimal possible with maximal drilling speed.

It is well known that for vertical motion of particles in the environment, they are always oriented with their long axes along the trajectory of motion. Such arrangement of particles moving with the vertical flow is the most stable, because the particles subjected to the action of gravitation forces and to the dynamic influence of the flow show the least resistance to the ambient flow. From this it follows that the sizes of pulp pipes in ED drilling of wells must be matched with sizes b and h of the largest drill cutting particles.

A sufficient matching condition must take into account the content of the largest drill cutting particles in the volume, that is, the probability of simultaneous occurrence of more than one largest particle in the pulp pipe cross section. In general, the most probable location of particles in the flow of the flushing out agent is such in which no more than one particle is contained in each pulp pipe cross section, since in this case they show the least resistance to the ambient flow. However, simultaneous occurrence of two and more particles in one pulp pipe cross section is inevitable, because drill cutting particles move with different velocities due to their different hydraulic sizes. Thus, it is possible to consider two cases of mutual location of particles:

- Particles are oriented with sizes b along the well radius,
- Particles are oriented in the above-indicated direction with sizes h_{max}.

Under these conditions, the most stable is the arrangement of particles oriented with size h_{max} along the radius. Hence, in the direct flushing scheme, the geometrical pulp pipe parameters must be matched with the size h_{max} of the largest particles.

Thus, the most important problem in determining the optimal flushing power is the estimation of a minimal possible flow rate of the flushing-out agent through the pulp pipe.

In the drill cutting flow comprising particles with sizes from dust ones to l_{max}, their velocities in the flushing agent flow become inversely proportional to their hydraulic sizes. Because of this, mutual displacement of particles is observed, their total flow resistance changes, and the velocity of motion of each particle, even for constant discharge rate Q_0, becomes variable.

For the direct flushing scheme, the pulp pipe is formed by the channel with ring cross section; one its surface (the wall of the well), as indicated above, has highly developed roughness. The second surface is formed by the steel pipe. For practical purposes, of interest is calculation of the equivalent roughness of this system. All this refers also to the channel of the down flow in the reverse flushing scheme.

The material consumption for flushing is estimated as the ratio of the flushing out agent lost during well drilling to the well length or as the ratio of the intensity of flushing out agent loss to the speed of well drilling. To almost irreplaceable losses, the portion of the flushing out agent belong that:

(a) is lost for leaks (absorption) in pores and cracks of the massive being drilled,
(b) is carried away by drill cutting particles in pores and to the surface due to wetting,
(c) is remained in the well after pumping out.

Thus, to achieve the minimal possible energy consumption for flushing in ED drilling, it is necessary to adjust the critical volume concentration of the consumed material to the pulp pipe diameter. Under this condition, the minimal possible energy consumption for flushing will change directly proportional to the squared radius of the drilling head.

Results of investigations of the main factors determining the efficiency of flushing in ED drilling can be generalized in the following theses:

1. The design parameters of the drilling rig determining the flushing regimes (rate of flow of the flushing agent through the pulp pipe and the volume concentration of the consumed solid material/drill cutting in hydromixture) are the electrode spacing in the drill head, cross section of the pulp pipe, diameter of the head, and diameter of pipes of the drilling rig.

The average volume productivity per impulse, impulse repetition frequency, and properties of rock and flushing out agent are among the factors influencing the flushing conditions.

2. The cross sectional sizes of the pulp are determined by the sizes of the largest drill cutting particles that depend on the electrode spacing in the drill head and the properties of rock. With increasing electrode spacing, the sizes of the largest particles increase with a decreasing speed.
3. The rate of flow of the agent through the pulp pipe depends on the electrode spacing. This dependence involves the average diameter of drill cutting particles, cross sectional area of the pulp pipe, and its hydraulic radius. With increasing electrode spacing, the average particle size increases with a decreasing intensity.

4. The required rate of the flushing out agent flow through the pulp pipe depends on the average volume impulse productivity and impulse repetition frequency. This dependence involves the volume solid material concentration in the discharge flow of the hydromixture.
5. To estimate the initial losses of the flow pressure due to the difference between the relative densities of the hydromixture and the flushing out agent, it is necessary to consider the actual volume concentration of the solid material in the pulp pipe, because it can differ significantly from the discharge volume concentration.
6. When the electrode spacing in the drill head is close to its radius (the condition of optimal efficiency of ED destruction), it is recommended to apply internal flushing for cleaning of wells of increased diameter. The volume discharge concentration of solid material in hydromixture should be matched with the pulp pipe geometry. Given that these conditions are satisfied, the minimal possible energy consumption for flushing will change directly proportional to the squared radius of the head.

7.3.5 Productivity and Energy Consumption

The optimal conditions of ED drilling for invariable parameters of the discharge circuit were investigated by changing the impulse voltage amplitude at constant discharge IVG capacity or the discharge capacity for constant IVG voltage amplitude. A change in the length S of the discharge gap (a change in the discharge circuit parameter) requires a search for optimal values in the family of curves with different S. In [1, 10] this dependence was obtained for microquartzites. It can be written as follows:

$$W = 90 \cdot S^{1.6}, \tag{7.4}$$

where $S \leq 7$ cm is the electrode spacing, in cm, and W is the energy, in J. Table 7.5 presents energies stored in the IVG that were calculated from Eq. (7.4) and the deposited energy for the indicated electrode spacing at which ED drilling was performed. Optimization of the deposited energy with minimal energy consumption for different S values led to the increase of the energy deposited into the discharge channel by ~3 times when S increased 6 times.

In Chap. 5 it was demonstrated that the increased energy deposition caused an increase in the destruction productivity, that is, promotes an increase in the drilling speed at $S = \text{const}$. Hence, the drilling speed can increase with S not only due to the

Table 7.5 Dependence of the energy stored in the IVG (calculated by Eq. (7.4)) on S

S, cm	20	40	60	80	100	120
W, J	272.8	827.1	1582.3	2507.2	3583.0	4796.6
W_S, J/mm	13.6	20.7	26.4	31.3	35.8	40.0

increased destruction volume with increasing S, but also due to the increasing energy deposition if the drilling process is optimized to minimize the power consumption. It is better to analyze the drilling process based on the energy deposition. For example, for identical values of energy deposition W_s = const as functions of the electrode spacing. Thus, a question inevitably arises: for what M value the drilling speed will be maximal?

The current state of knowledge on ED drilling does not allow a dependence of the productivity on the two parameters—electrode spacing and energy deposition—to be determined, because a change of one of them causes the change of time and character of energy deposition into the discharge channel, which inevitably affects the destruction productivity. Analogous difficulties arise when analyzing the results of destruction of rock samples. In this regard, it was suggested to consider the energy characteristics of the ED destruction and drilling processes taking into account the energy deposition. Unfortunately, such approach can be realized not always. For example, when drilling in the field conditions, it is extremely inconvenient to fulfill the requirements of identical energy deposition for different S values. This required a simplified, but sufficiently effective approach to a solution of the given problem to be used.

Based on generalization of the available experimental data, the dependences of the drilling productivity of granite and microquartzite were obtained for one impulse (curve 1) and of energy consumption (curve 2) on the electrode spacing for different drilling rigs, Fig. 7.18.

The character of changing the dependence of the drilling productivity on S is analogous to that obtained for destruction of rocks with a two-electrode system. The dependence $Q = f(S)$ in Fig. 7.18 can be described by the empirical expression of the type

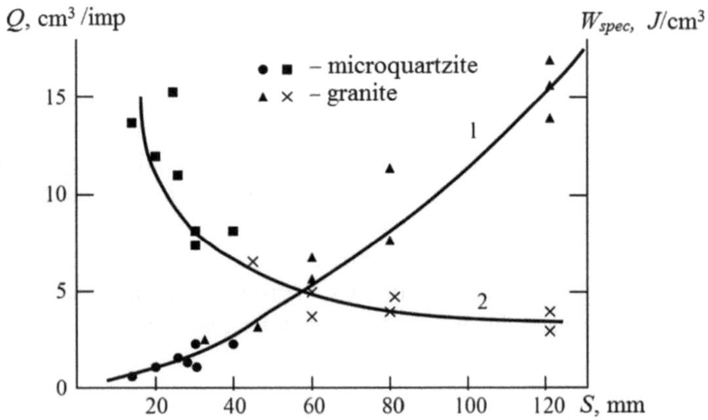

Fig. 7.18 Dependence of the productivity (1) and energy consumption (2) in rock drilling on the electrode spacing for drill heads with different diameters. Here curves 1 and 2 are for results of calculation from Eqs. (7.6) and (7.2), respectively

Fig. 7.19 Dependence of the drilling productivity on the energy deposition for microquartzite at $S = 20$ (curve 1), 30 (curve 2), 40 (curve 3), and 50 mm (curve 4)

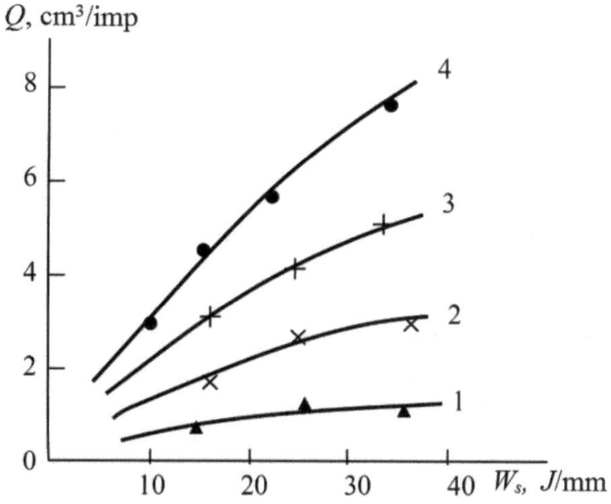

$$Q = Q_0 \cdot (S_i / S_0)^b, \quad (7.5)$$

where $Q_0 = 1.0$ cm^3/impulses is the drilling productivity for $S_0 = 2.0$ cm, and $b = 1.55$ is the experimental coefficient. In Chap. 5 analogous dependence was described by Eq. (5.3), where the exponent $a = 2.0$–2.15, that is, a faster increase in the productivity was observed in the laboratory conditions than in actual drilling. This effect is caused by specific conditions of rock destruction by drilling (see below).

With increasing electrode spacing, the energy consumption quickly decreases, especially up to $S = 40$–50 mm (curve 2 in Fig. 7.18). Analogous character of the dependence $W_{spec} = f(S)$ was also observed for the destruction of rocks with a two-electrode system. However, the empirical expression was slightly different:

$$W_{spec} = W_0 \cdot (S_i / S_0)^{-c}, \quad (7.6)$$

where $W_0 = 1200$ J/cm for $S = 2.0$ cm and $c = 0.8$ is the experimental coefficient. Dependences (7.5) and (7.6) convincingly show prospects for ED drilling of wells of large diameter with correspondingly large electrode spacing. Both dependences can be used for preliminary estimation of the drilling characteristics for preset (chosen) electrode spacing.

In Chap. 5 it was shown that the productivity of rock destruction with a two-electrode system depends not only on the electrode spacing, but also on the energy deposition. This regularity should be tested for actual drilling. To this end, from a sufficiently large volume of accumulated data of drilling, results obtained for concrete rock and close regimes of energy deposition were chosen.

Figure 7.19 shows the dependences of the drilling productivity on the energy deposition for microquartzite and indicated electrode spacing. The productivity increases with energy deposition the faster, the larger is the electrode spacing. In this case, the dependence $Q = f(M)$ is saturated as for the two-electrode system.

It should be noted that the transition to saturation is displaced toward larger energy deposition with increasing S. This character of changes allows the following conclusions to be drawn:

(1) Saturation of the dependence $Q = f(W_s)$ for S = const demonstrates that the drilling productivity (rate) will increase with energy deposition until the saturation; further increase in W_s will not practically increase the productivity (rate) of drilling, which is primarily determined by the dynamics of crack growth when the discharge channel penetrates into the rock and by the limited spacing; this is clearly seen in Fig. 7.19;
(2) The dependences shown in Fig. 7.19 allow us to suggest that for the preset electrode spacing, the optimal amount of energy deposition at which the drilling productivity (rate) is maximal is observed. Thus, an increase in the productivity (rate) of ED drillings for the chosen electrode spacing has the physical restriction caused by the amount of energy deposition into the discharge channel;
(3) A faster increase in the drilling productivity with electrode spacing at constant W_s is yet another demonstration of good prospects for drilling of wells with large diameter with increased electrode spacing.

The dependences of the drilling productivity on the energy deposition were also obtained in drilling of wells with large diameter in granite using a BI-325 drilling rig with DH-360-80 and DH-360-120 drill heads. The energy stored in the IVG changed mainly due to the change of the discharge capacity and partly the of discharge voltage. In addition, the liquid flow rate increased from 1900 to 3900 L/min.

Figure 7.20 shows results of drilling of the well 400 mm in diameter.

An increase in the energy deposition at $S = 80$ and 120 mm leads to the increase of the drilling productivity that is faster for $S = 120$ mm. Thus, for $S = 80$ mm, an increase in the energy deposition by a factor of 2.4 caused the productivity to increase by a factor of 2.35, and for $S = 120$ mm, the productivity increased by a factor of 1.74 caused the increase of the productivity by a factor of 2.4. Here the productivity curve is not yet saturated for the indicated S values, though the energy stored in the IVG significantly (by a factor of 1.98 at $S = 80$ mm) exceeded that presented in Table 7.5. Proceeding from the dependences shown in Fig. 7.19, the saturation of the dependences shown in Fig. 7.20 should be observed at much larger energy deposition than that realized in these experiments.

A comparison of the productivity of ED drilling and ED destruction with a two-electrode system depending on S is illustrated by Fig. 7.21. Here curve 2 shows the results for the two-electrode system. Since the destruction productivity depends both on the electrode spacing and energy deposition, the comparison is performed for W_s = const.

Results for the two-electrode system were obtained upon exposure to three impulses, when the destruction productivity is maximal (see Sect. 5.1 of Chap. 5). The productivity of the two-electrode system for the entire range of electrode spacing was by a factor of 1.4–1.9 higher than in drilling with multielectrode head (curve 1).

Fig. 7.20 Dependence of the productivity of granite drilling on the energy deposition at S = 80 (curve 1) and 120 mm (curve 2)

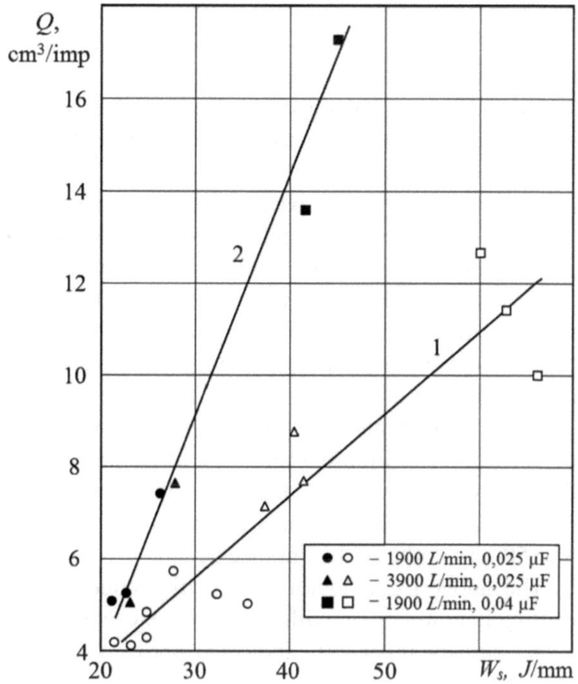

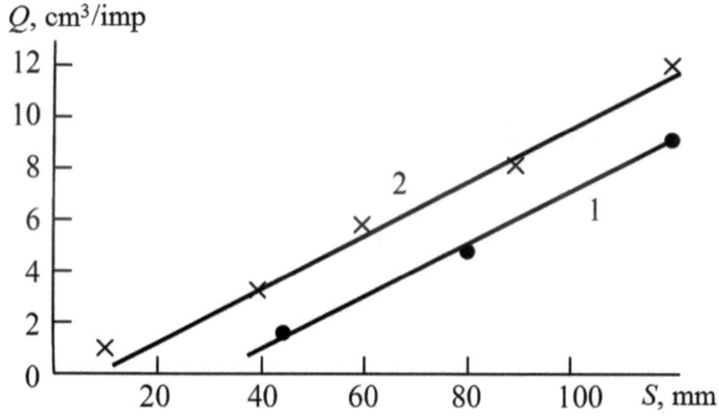

Fig. 7.21 Dependence of the drilling productivity on the electrode spacing at W_S = const. Here curve 1 is for W_S = 20.1–23.4 J/mm and curve 2 is for W_S = 21.7–25.3 J/mm

Calculations of the actual drilling speed based on the results of experiments with the two-electrode system require refinement owing to the specificity of conditions at the well bottom.

First, as indicated above, the drilling process is cyclic in character. The cycle is terminated when the rock has been destructed in all elementary two-electrode discharge gaps. It is necessary to act on each two-electrode gap of the drill head by impulses whose number is much greater than three (ten and more), which reduces the productivity recalculated per impulse.

Second, not each voltage impulse applied to the well bottom caused effective rock destruction; in this connection, the concept of "impulse utilization coefficient" was introduced. This coefficient changes from 0.75 to 0.96 depending on the voltage and increases with electrode spacing. This effect also decreases the drilling productivity.

Third, in connection with high roughness of the bottom and walls of the well, the discharge often develops in the combined system rock—flushing out liquid, and a portion of the discharge energy is released in the liquid rather than in the rock, thereby reducing the destruction efficiency. The roughness of walls of the well can be judged from the cavernogram recorded with a two-laser caliper in drilling of granite in the test site near Bergen (Norway), Fig. 7.22. The average well diameter equal to 400 mm decreased with depth to 390 mm, which was due to granite erosion in the upper part of the well. The minimal diameter of protrusions was 376 mm, and the maximal diameter of gaps was 400 mm, that is, the maximal difference between the diameters for the side was 17 mm that was close to the discharge channel penetration depth at $S = 50$ mm under multi-impulse action.

The above-listed factors inevitably reduce the productivity of destruction by drilling in comparison with the laboratory experiments with two-electrode system.

As already repeatedly indicated above, the energy consumption is one of the important characteristics of rock drilling in any method. In this connection, it is necessary to consider the influence of the energy deposition on the energy consumption for wells with large diameter in ED drilling. In a number of experimental works it was demonstrated that the energy deposition is minimal for a certain energy stored in the IVG. From these dependences it follows that with increasing electrode spacing, the minimal energy consumption is displaced toward larger values of energies stored in the IVG, and the absolute value of energy consumption decreases. Analogous dependences are shown in Fig. 7.23 illustrating well drilling using a BI-325 rig with electrode spacing of 80 and 120 mm, respectively, for deposition energy in the range 24–61 J.

At $S = 80$ mm, a minimal energy consumption is observed at $W_s = 40$ J/mm, and at $S = 120$ mm, it is displaced toward $W_s = 45$–50 J/mm. Thus, the minimal W_{spec} value at $S = 120$ mm decreases by a factor of 1.25. These dependences confirm the general regularity in changing of the energy consumption as functions of the energy stored in the IVG that was indicated above.

The experimental values of the energy deposition for minimal energy consumptions at $S = 80$ and 120 mm are in good agreement with the results presented in Table 5.3 that were calculated from Eq. (5.3) with approximation for $S > 70$ mm. With further increase in the energy deposition, the energy consumption increases; in this case, the productivity also increases (Fig. 5.7). Hence, the optimization of ED drilling with respect to the minimal energy consumption is a special case that can be realized in concrete drilling conditions.

Fig. 7.22 Cavernogram of a well in granite

7.3.6 *Drilling Speed*

To analyze a dependence of the drilling speed on the electrode spacing, it is necessary to limit the change of the energy deposition and to consider drilling of wells with a concrete diameter of the drill head. Figure 7.24 shows the dependences of the speed of drilling of the indicated rocks on the electrode spacing for drill heads of different diameters.

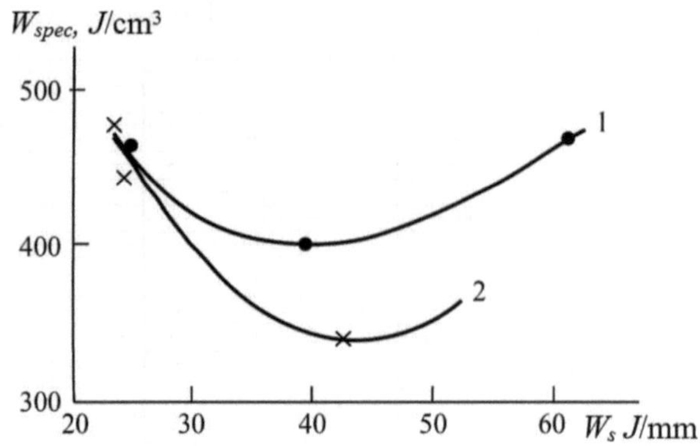

Fig. 7.23 Dependences of the energy consumption on the energy deposition in granite drilling with a BI-325 drilling rig at S = 80 (curve 1) and 120 mm (curve 2)

In this case, the drilling conditions for a concrete drill head were chosen with close values of the energy deposition at different S: for microquartzite at D_{head} = 120 mm, the energy deposition changed in the range 3–40 J/mm, and at D_{head} = 150 mm, it changed in the range 20–27 J/mm; for sandstone at D_{head} = 140 mm, the deposited energy changed in the range 31.6–32.5 J/mm; and for permafrost at D_{head} = 200 mm, it changed in the range 28.1–28.9 J/mm. In the examined range, an increase in the electrode spacing leads to almost linear increase of the drilling speed for heads of all considered diameters. The increment of the speed with increasing spacing was greater for less strong rocks (curves 3 and 4).

An increase in the drill head diameter at S = const causes the drilling rate to decrease (curves 1 and 2 in Fig. 7.19). Curve 1 in Fig. 7.24 shows the dependence of the speed of microquartzite drilling on the drill head diameter at S = 20 mm.

Despite a significant difference in the deposited energy (18.1 J/mm for D_{head} = 90 mm and 38.4 J/mm for D_{head} = 120 mm), an increase in D_{head} by a factor of 1.66 caused a decrease in the drilling speed by a factor of 2.1, since for close values of the drilling productivity at S = 20 mm in the indicated range of W_s values the

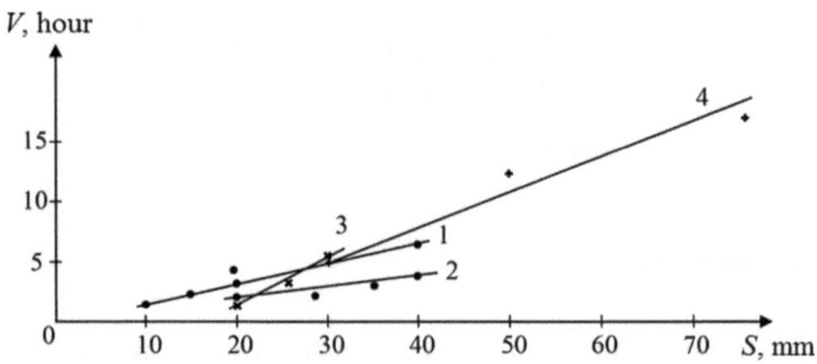

Fig. 7.24 Dependence of the drilling speed on the electrode spacing in drilling of microquartzite with D_{head} = 120 (curve 1) and 150 mm (curve 2), sandstone with D_{head} = 140 mm (curve 3), and permafrost with D_{head} = 200 mm (curve 4)

destruction of a larger area at the well bottom requires longer time. This effect is confirmed in drilling of granite using heads with diameters $D_{head} = 120$ mm ($W_s = 34.2$ J/mm) and 360 mm ($W_s = 27.9$ J/mm) at $S = 40$ mm. The drilling speed was 6 m/h and 2.3 m/h, respectively, that is, it decreased by a factor of 2.6 when the diameter increased by 3 times. Hence, to retain or to increase the drilling speed with increasing head diameter, the electrode spacing must be increased. In this case, optimization with respect to the energy deposition is also required.

The diameter of the drill head limits the maximum possible electrode spacing that for the conventional radially-tangential drill with rod electrodes does not exceed the head radius. The actual distances are always smaller, since the electrodes have a certain area of contact with the well bottom. The minimal number of electrodes is seven. For the preset electrode spacing, the number of electrodes at the well bottom can be different because of their different configurations and sizes. In this regard, the resultant drilling speed for various head diameters must be compared for the same numbers of electrodes at the well bottom (including some assumptions concerning the influence of the electrode area on the destruction efficiency).

We processed the data of field drilling with heads of different diameters but with the same number of electrodes equal to 11. (Unfortunately, they were obtained in drilling of different rocks.) In this case, the energy deposition changed in the range 25.5–35.0 J/mm. Curve 2 in Fig. 7.25 shows the dependence of the drilling speed on the drill head diameter at the indicated electrode spacing. It can be seen that the drilling speed increases with drill head diameter provided that in this case, the electrode spacing correspondingly increases. For example, for microquartzite and granite with relatively close ED drilling characteristics, an increase in the head diameter by a factor of 2.45 (from 110 to 270 mm) and in the electrode spacing by a factor of 4.66 (from 15 to 70 mm) caused the increase of the drilling speed by a factor of 4.1. The optimization of the electrode spacing when the number of electrodes at the well bottom is equal to 7 will increase the drilling speed even more. It is obvious that the electrode spacing causes an increase in the operating voltage and energy stored in the IVG as well as an increase in the productivity of the flushing system intended for removal of drill cuttings, because the volume of the destructed rock and the relative fraction of large drill cuttings increase.

As demonstrated above, the productivity of ED drilling increases with energy deposition in the examined range of M values, and the faster, the larger is the electrode spacing (Fig. 7.26). This inevitably leads to an increase in the drilling speed. Figure 7.26 shows the dependence of the drilling speed on the energy deposition in drilling of wells with drill heads having the diameter $D_{head} = 360$ mm for different values of the electrode spacing. In this case, the energy deposition for each S value was changed by changing the discharge IVG capacity—the most effective factor of the rock ED destruction at an optimal operating voltage. The character of changes of the drilling speed with energy deposition is similar to that of the drilling productivity (Fig. 7.20). The drilling speed linearly increases with the energy deposition in the examined range at both values of the electrode spacing, but at $S = 120$ mm, the increase of the drilling speed is faster than at $S = 80$ mm. Thus, in the range $W_s =$

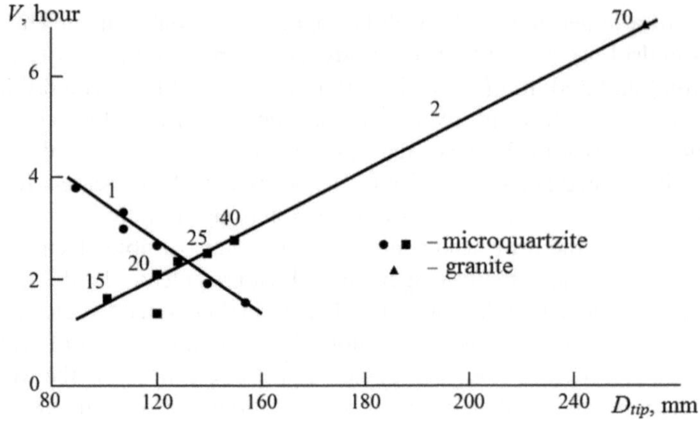

Fig. 7.25 Dependences of the drilling speed on the diameter of the drill head at $S = 20$ mm (curve 1) and at different S values (in mm) indicated near the symbols (curve 2)

Fig. 7.26 Dependences of the drilling speed on the energy deposition at $S = 80$ (curve 1), 120 (curve 2), and 50 mm (curve 3)

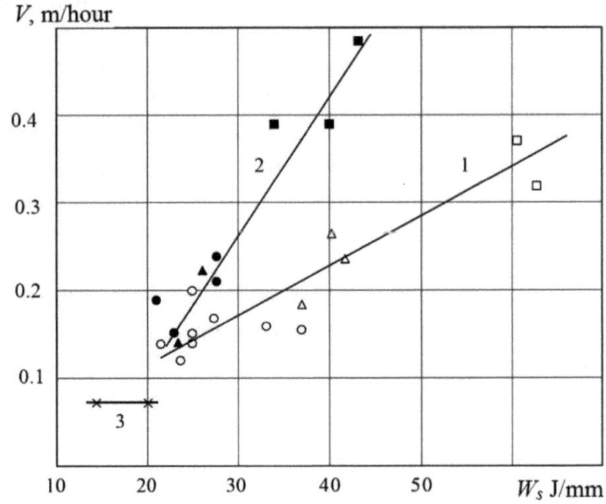

23–40 J/mm, the drilling speed at $S = 80$ mm increased by a factor of 1.76, whereas at $S = 120$ mm, it increased by a factor of 2.66.

Results of investigations of the dependence of the ED drilling speed on different factors demonstrated that the drilling efficiency increased with electrode spacing, well diameter, energy deposited into the discharge channel when the optimal flushing regimes are provided.

References

1. Maurer WC (1968) Novell drilling techniques. Pergamon Press, QB, 110 p
2. Vajov VF, Datskevich SY, Zhurkov MY et al (2012) Grain size composition and distribution in drill cuttings in the electric impulse destruction of rocks. J Min Sci 48(1):103–109
3. Siomkin BV, Usov AF, Kuretz VI (1995) Bases of electropulse destruction of materials. St. Petersburg, Nauka, 276 p (in Russian)
4. Protasov YI (2001) Destruction of rocks. Publishing house MGGU, Moscow, 453 p (in Russian)
5. Kuchinsky GS (1973) High-voltage pulse capacitors. Publishing house Energiya, Leningrad, 175 p (in Russian)
6. Ushakov VY, Dulzon AA (2001) Performance capability of technological installations using the electrical discharge or conductor electrical explosion energy. In: International conference on pulsed power applications. Gelsenkirchen, 27–29 Mar 2001, Pap. Numb. A.03
7. Kuretz VI, Usov AF, Zuckerman VA (2002) Electropulse disintegration of materials. Publishing House of the Kola Scientific Center of the Russian Academy of Sciences, Apatity, 324 p (in Russian)
8. Vorob'ev AA, Vorob'ev GA, Chepikov AT (1998) Certificate of the opening of NA-122. Regularities of breakdown of a solid dielectric at the interface with a liquid dielectric under the action of a voltage pulse, from 29.04.1998 with the priority of 14.12.1961 (in Russian)
9. Smolyanytsky AA, Genkin EE (1974) Investigation of the nature of the destruction of rocks during an electric explosion. Gorny J Izvestya vuzov (9):69–71 (in Russian)
10. Boev S, Vajov V, Jgun D et al (1997) Electropulse technology of material distraction and boring. In: IEEE international pulsed power conference, USA, Baltimore, pp 220–225

Chapter 8
Destruction and Recycling of Reinforced Concrete Products

8.1 Urgency of the Problem and Ways of Its Solution

An increase in the volume of concrete and reinforced concrete production and requirements to their quality leads to accumulation of wastes in all stages of construction, including the precast concrete industry. Tens of millions of cubic meters of sub-standard products and designs have been accumulated in big cities. The main volume of sub-standard products remains in enterprises-manufactures, polluting their territories.

Sources of wastes are also reinforced concrete products (RCPs) from buildings and structures being demolished, disassembled, or reconstructed, road coatings destroyed by earthquakes, mudslides, and others natural phenomena and hostilities and terrorist acts.

Until recently, reinforced concrete wastes have almost unused because there were no economically suitable methods of their recycling and no equipment for destruction of reinforced concrete products and structures. And this is despite the fact that crushed concrete can be recycled to prepare non-essential reinforced concrete products, whereas the RCP armature can be used as a scrap metal; under certain conditions, it can even be recycled directly. Alienation of plots of land is simultaneously prevented, and other positive effects are achieved.

For reinforced concrete structures, the first stage—deconstruction (extraction of iron armature)—is the most difficult and energy-intensive. At the beginning of the "reinforced concrete era," concrete building structures were destructed with a steel wracking ball or by continuous impact destruction with a large crusher. However, these methods cause the undesirable phenomena: large energy consumption, noise, vibration, and dust. Their application in cities is harmful for citizens and becomes a social problem. The explosive destruction of building concrete structures is widely used in European countries. In Japan, hydraulic crushers that produce less noise, vibration, and dust in comparison with other destruction methods are most widespread. In partial disassembly of buildings, when the application of conven-

V. Y. Ushakov et al., *Electro-discharge Technology for Drilling Wells and Concrete Destruction*, https://doi.org/10.1007/978-3-030-04591-3_8

tional methods is complicated due to the presence of people living in them, it is required to perform disassembly with observance of special security measures with low level of vibration, noise, and dust. In this case, concrete structures are destructed using heating, diamond saws, or abrasive water jets.

Sorting and warehousing of parts of reinforced concrete structures and their transportation to recycling sites do not require special machines, since conventional mechanisms: bulldozers, elevating cranes, cutting torches, etc. allow these problems to be solved.

The technology of RCP destruction (primary crushing) in many countries is based on the application of the crushing-sorting equipment used for stone recycling. However, the presence of armature in RCPs tightens the requirements imposed on installations and causes a number of drawbacks:

- Large metal consumption
- Considerable dimensions
- Low operational efficiency
- Presence of dust produced during operation of the installation
- Bad quality of crushed concrete because of armature reminders
- Additional expenses for armature crushing and cutting.

The application of rotary-hammer and jaw crushers for primary crushing of substandard RCPs is limited by their sizes: 2–3 m long and 1 m wide, which requires preliminary destruction of large-size RCPs down to allowable sizes.

The electrohydraulic method with preliminary borehole penetration is effective for destruction of small RCPs. For $U_0 \sim (5\text{–}40) \times 10^3$ V, $W_0 \sim 10^5$ J, and impulse repetition period of 1 min, the productivity reaches (100–120) cm^3/impulse.

The complexity of the problem of reinforced concrete destruction for recycling of all components without exception is that concrete must be separated completely from the armature. Under this condition, the metal can be remelted and rubble can be recycled. Most of the above-listed methods are aimed at destruction of RCPs into fragments with sizes that allow one to use them for piling, preparation for concreting, or transportation to dumping sites. In the first case, the potential of waste recycling is incompletely realized, and in the second case, money are spent for waste transportation to damping sites and plots of land are littered. As already indicated above, the electrodischarge technology can be wasteless, and the armature network of structures together with imbedded parts can be used directly without preliminary preparation.

In the first works devoted to research and development of the ED technology of concrete and RCP destruction, it was established that voltage of hundreds of kilovolts and highly intensive regimes of energy liberation in the discharge channel, including discharge currents of tens of kiloamperes, instantaneous powers up to $10^9\text{–}10^{10}$ W, and pressures up to 10^{10} Pa, must be used due to high impulse electric strength of concrete and large RCP thickness. The basic components of the equipment, including capacitive storage devices, switchers, and electrode systems under the action of such intensive factors, as a rule, have limited lifetime, require high depreciation charges, and are unreliable in operation. To make this technology competitive with traditional technologies of RCP destruction, researchers are required focused on a study of separate types of energy consumption in a wide spectrum of energy regimes to provide the possibility of EDT application in low-intensive discharge regimes facilitating the working conditions of the equipment and, as a consequence, increasing the reliability and the lifetime of the equipment. The problem is complicated by the fact that RCPs are multicomponent materials that differ by the degree of saturation with the armature and the content of the filler of different fractional composition depending on the product type. This requires knowledge of technical and economic indicators of recycling of concrete product types and search for optimal destruction regimes to extract for recycling the concrete filler.

8.2 Studies of the Characteristics of ED Concrete Destruction on Samples

8.2.1 Experimental Procedure

Investigations were performed for concrete mark 200 samples with sizes of $10 \times 10 \times 10$ cm for 11 energy liberation regimes, Table 8.1.

The regime of energy liberation in the discharge channel was changed by changing the inductance L of the discharge circuit and the voltage impulse amplitude U_0. Breakdown of concrete samples by single voltage impulses was initiated in transformer oil using the "core-plane" system of electrodes. The energy W_Σ liberated in the discharge channel during time t of the discharge current flow and the energy W_1 of the first half-period were determined from the current waveforms and the voltage drop across the spark channel. For each combination of experimental conditions, breakdowns were initiated in 8–10 samples. The external view of the destructed samples for the indicated regimes of energy deposition is shown in Fig. 8.1.

After destruction of concrete and evaporation of transformer oil from sample fragments, sieve analysis was performed using standard set of sieves of wear-resistant from the results of which the average fragment size characteristics were determined (Table 8.2).

These size characteristics were used to determine the surface of concrete fragments S_{scf} and their average size D. To calculate S_{scf}, we used the formula for the elementary size class

$$dS_{nfs} = \frac{6}{x\rho_0} dP, \tag{8.1}$$

where dP is the weight yield of the elementary class, x is the average size of the class, and ρ_0 is the concrete density.

The average sizes of concrete fragments were calculated from the formula

Table 8.1 Experimental conditions of destruction of concrete samples

Regime	U_0, kV	L, μG	$T_{o.p}$, μs	W_0, J	Average values				
					N_{max}, MW	W_Σ, J	W_1, J	t, μs	Efficiency of the discharge circuit, η, %
1	150	13.6	5.6	656	199.7	460.3	359.5	8.4	70.2
2	192	13.6	5.6	1080	336.9	743.9	528.0	11.2	68.9
3	240	13.6	5.6	1680	368.7	963.5	343.5	16.8	57.4
4	300	13.6	5.6	2625	469.1	1476.4	848.8	16.8	56.2
5	360	13.6	5.6	3780	847.0	2250.7	1362.6	22.4	59.5
6	408	13.6	5.6	4855	1040.0	2521.9	1394.1	25.2	51.9
7	444	13.6	5.6	5750	1049.8	2701.8	1352.5	25.2	47.0
8	360	63.0	12.0	3780	318.1	2283.4	1187.7	60.0	59.9
9	360	210.0	22.0	3780	169.3	2337.9	822.7	121.0	61.8
10	360	840.0	44.0	3780	54.93	2218.9	743.8	396.0	58.7
11	360	2650.0	78.0	3780	31.6	1709.7	672.5	546.0	45.2

Fig. 8.1 External view of destructed samples. The number of the sample corresponds to the regime of energy deposition in Table 8.1

Table 8.2 Size distribution of the destructed samples

Regime	Weight content of fractions, %					
	(0–5) mm	(5–10) mm	(10–20) mm	(20–40) mm	(40–70) mm	(70–100) mm
1	0.13	0.14	0.35	0.53	0.00	98.85
2	0.24	0.30	1.04	2.00	12.15	84.27
3	0.36	0.41	1.60	2.99	24.75	69.89
4	0.59	0.75	2.68	5.28	28.44	62.26
5	1.13	1.17	4.12	7.28	32.76	53.54
6	0.79	1.59	6.12	11.12	80.38	0.00
7	1.12	1.35	5.17	11.69	80.68	0.00
8	1.05	1.18	3.25	6.16	22.00	66.36
9	0.71	0.81	2.86	7.08	21.45	67.09
10	1.00	0.74	2.57	6.52	19.58	69.59
11	0.85	0.73	2.49	4.1	21.13	70.70

$$D = \frac{\Sigma dP}{\Sigma \frac{dP}{x}}. \tag{8.2}$$

8.2.2 *Influence of the Regime of Energy Liberation in the Discharge Channel on the Destruction Characteristics*

The influence of energy W_0 stored in the capacitor and energy liberated in the breakdown channel on the specific energy consumption W_{spec} at destruction of samples (regimes 1–7 in Table 8.1) as well as the influence of the period of energy deposition on W_{spec} (regimes 5 and 8–11 in Table 8.1) were investigated. The W_{spec} values for each regime were calculated as the arithmetic mean for a series of experiments.

In the first cycle of research, the energy was varied by changing the voltage U_0 given that the discharge capacitance (0.0583 μF) and the inductance (14 μH) remained unchanged. The voltage increase from 150 to 445 kV was accompanied by the growth of the peak pressure in the breakdown channel from 9.5×10^8 to 16.3×10^8 Pa. In this case, the period of energy deposition increased from 8.4 to 25.2 μs, that is, the regime of energy deposition was displaced from the aperiodicity boundary toward oscillatory regimes. The estimation of this displacement by the similarity criterion Π demonstrated its decrease from 0.212 to 0.072.

Figure 8.2 shows the dependences $W_{spec} = f(W_\Sigma)$ for fractions of 5–10, 10–20, 20–40, and 5–70 mm, respectively. The choice of these size classes is explained by the fact that the filler of these fractions was used for concrete preparation. The change of the area of newly formed surface S_{scf} chosen for integral estimation of the destruction and the specific energy consumed for the formation of the unit surface ω_s depending on W_Σ are shown in Fig. 8.3. From the figure it can be seen that the minimal specific energy consumption for all fractions corresponds to regime 6 in which fragments with sizes exceeding 70 mm were not observed and W_{spec} for the fraction 5–70 mm was 1.1 J/g with specific energy consumption $W_s = 2$ J/cm^2. This regime was characterized by discharge current of 17.8 kA and instantaneous capacitance of 1.04×10^9 W.

The maximal pressure in the spark channel was 15.7×10^8 Pa. Attention is attracted to the fact that the specific energy consumption changes also insignificantly for regime 3 for which $W_s = 2.1$ J/cm^2 when W_Σ increases from 800 to 2500 J and makes $W_{spec} = 1.3$ J/g (fraction 5–70 mm), the discharge current decreased by 36% in comparison with the optimal regime (regime 6), instantaneous capacitances decreased by 65%, peak pressures by 23%, and discharge circuit efficiency increased by 10%.

To obtain the same productivity in regime 3 (Table 8.1), the impulse frequency should be increased three times. To simplify the operating conditions of the equipment, the discharge currents must be limited by increasing the impedance of the discharge circuit.

In the second cycle of research, the energy deposition regime was changed by switching on air throttles with different inductances L at the voltage $U_0 = 360$ kV and capacitance $C_{disch} = 0.0583$ μF. The increase of the inductance from 14 to 2560 μF was accompanied by the increase of the energy deposition time from 25.2 to 540 μs and the decrease of the peak pressure from 15.7×10^8 to 0.95×10^8 Pa, the discharge

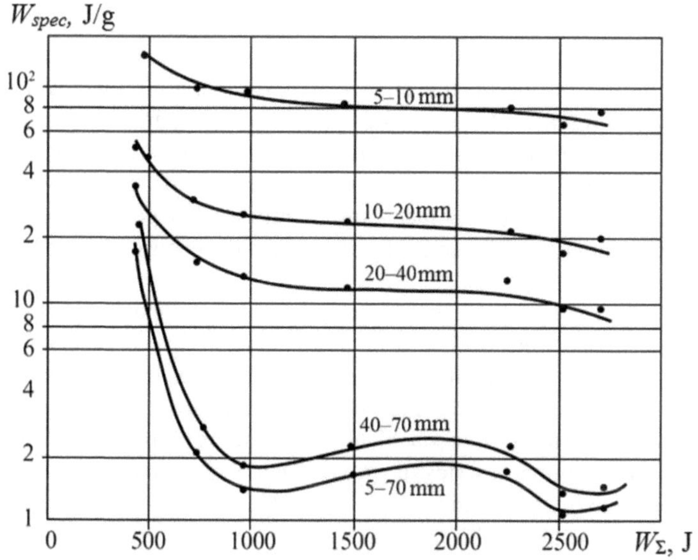

Fig. 8.2 Dependence of the specific energy consumption on W_Σ

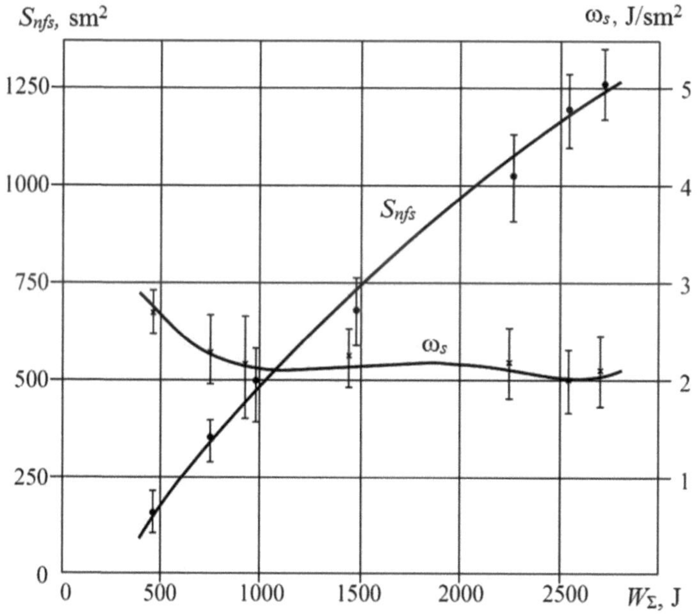

Fig. 8.3 Dependences of S_{scf} and ω_s on the energy consumption W_Σ

current amplitude from 15.2 to 1.45 kA, and the instantaneous power from 8.5×10^8 to 3.2×10^7 W.

Figures 8.4 and 8.5 illustrate the dependences of S_{scf}, W_{spec}, and ω_s on the discharge current oscillation period $T_{o.p}$ chosen to characterize numerically the energy deposition regime. Considering that change of the discharge circuit inductance simultaneously caused significant changes of a number of the parameters characterizing the dynamics of energy deposition and the interaction of generated stress waves with the sample boundaries, it is difficult to interpret the dependences shown in Figs. 8.4 and 8.5, including the interesting fact—sufficiently weak dependence of the specific energy consumption in a wide range of variations of the energy consumption regime.

The application of the regime most sparing for the equipment has led to an increase in the specific energy consumption by no more than 25% in comparison with regime No. 5. This suggests that in this regime, when the wavelength of the generated mechanical stresses significantly exceeds the sample sizes, it is dispersed before the termination of forming of mechanical perturbations by the breakdown channel. To verify this assumption, a series of experiments on concrete destruction was performed in regime No. 11 with the sample in a polyethylene sheathing preventing the dispersion of fragments. In this case, a more uniform fragmentation of large fractions without fragments whose sizes exceeded 70 mm was observed. The content of small fractions did not change significantly. The experiment has confirmed a definite role of perturbations generated by the breakdown channel in late stages of discharge current oscillations.

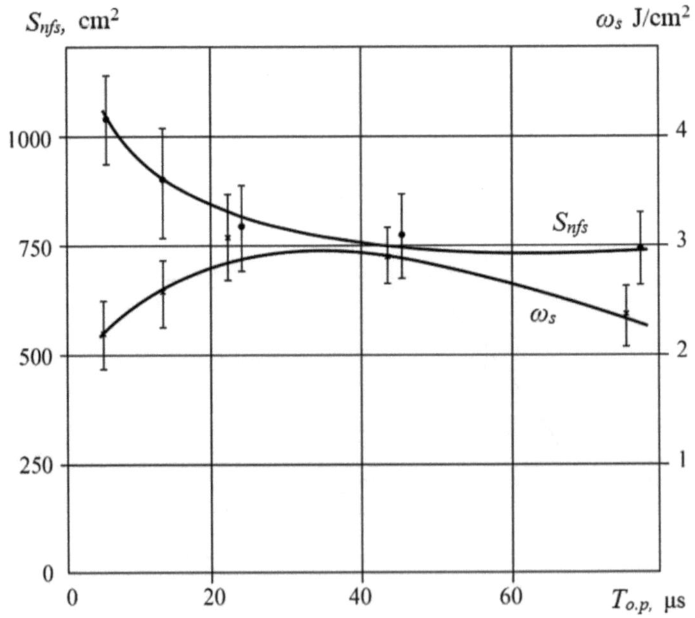

Fig. 8.4 Dependences of S_{scf} and ω_s on $T_{o.p}$

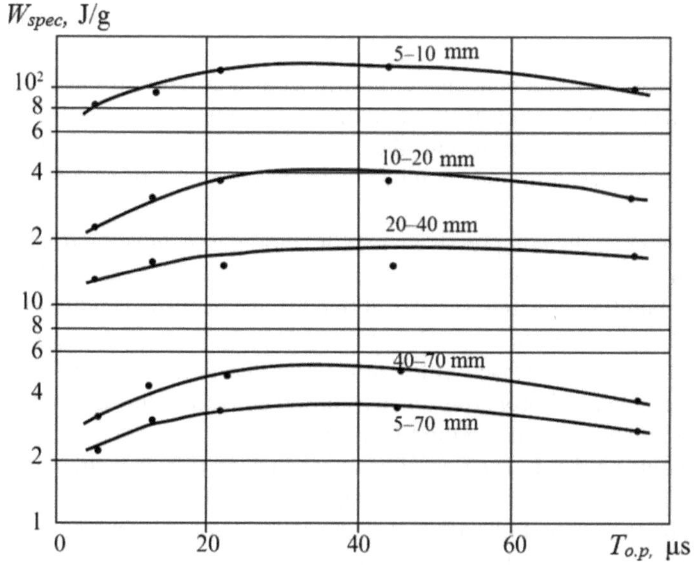

Fig. 8.5 Dependence of specific energy consumptions on $T_{o.p}$

That is, the application of regimes with low energy consumption and long discharge current duration for the destruction of large RCPs will have an additional effect in comparison with that observed in the experiments discussed above, since separation of fragments from the massive will be limited in the early stages by the product mass and the armature network.

8.2.3 Influence of Mechanical Strength of Concretes on the Effectiveness of Their Destruction

One of the factors influencing the effectiveness of destruction of solids, in addition to loading conditions, is their physicomechanical properties. To take into account the strength and elastic characteristics of materials, the parameters necessary for it can be found in the literature. To take into account the physicomechanical properties in destruction of rocks and artificial materials by impulse electric discharges, the relationship $\sigma_{demol}/\sqrt{E\rho_0}$ was used in [1], where σ_{demol} is the tensile strength and E is the elasticity modulus; moreover, the greater its value, the harder it is to destruct the material. In experiments with the same impulse parameters and sample geometry, granite for which $\sigma_{demol}/\sqrt{E\rho_0} \sim 0.4 \times 10^2$ was destructed more intensively than ceramics having $\sigma_{demol}/\sqrt{E\rho_0} \sim 1.5 \times 10^2$. For the discharge circuit parameters $U_0 = 237$ kV, $C_{disch} = 0.02$ μF, and $L = 38$ μH, ceramic samples 15 mm thick were destructed only in the crush zone, while the cracking zone was also formed in granite samples.

Some researchers accept acoustic rigidity λ_o as a criterion for estimation of the rock parameters. According to the authors of [2], the degree of electroexplosive destruction of rocks increases with the acoustic rigidity. The maximal destruction intensity was observed in rocks with λ_o in the range from 10^7 to 2.5×10^7 kg/m^3 m/s.

In experiments, the compressive strength being the main strength characteristic was chosen to consider physico-mechanical properties of concrete. For the most widespread building concrete (grades 200–600), the acoustic rigidity changed in a rather small interval from 8×10^6 to 11×10^6 kg/m^3 m/s. This suggests that the character of concrete destruction by an electric discharge will weakly depend on its physico-mechanical properties, as was observed earlier during crack and trench penetration into concrete using the ED technology.

Experiments on the destruction of concretes having different strengths attendant to changes in the dynamics of energy deposition in the breakdown channel were performed by the method of multifactor planning of the experiment. As criteria for estimation of the destructibility of concrete, the area S_{scf} of the formed fragments and their average size D determined from their size analysis were accepted. The following parameters and ranges of their changes were chosen:

Energy stored in the IVG $W_o = 625$–5900 J,
Discharge current period $T_{o.p} = 6$–80 μs,
Concrete grade $B_c = 100$–600.

Experiments were performed using a rotatable plan of the second order.

It was established that the area of the formed concrete fragments and their average size depended significantly on the impulse energy and somewhat weaker on impulse duration; it was not possible to establish the influence of the concrete grade on S_{scf} and D for the accepted significance level $\alpha = 0.05$. The results of the experiments confirmed the possibility of effective ED destruction of concrete having different grades.

8.3 Main Technological Characteristics of the ED Destruction of Sub-standard Reinforced Concrete Products from the Results of the Laboratory Installation Testing

Recycling of sub-standard reinforced concrete was carried out in two stages, each comprising a number of successive stages.

Stage I:

- Dismantling of building structures
- Primary sorting and warehousing

- Transportation of a concrete scrap to a reprocessing site.

Stage II (considered in Chap. 9):

- Reprocessing
- Recycling.

Results of investigations of RCP destruction to extract armature (dismantling of building structures) are presented below. In connection with the variety of reinforced concrete product types, different sizes, configurations, degrees of saturation by armature, and concrete grades, it is necessary to estimate the productivity of the installation and the specific energy consumption for the destruction of a concrete product. In addition, such investigations are necessary to expand the nomenclature of the above-listed products destructed using the ED technology.

Based on results of investigations of impulse breakdown voltage for concretes of different grades, the energy characteristics of the spark channel in concrete and the calculated parameters of mechanical stress waves generated by the breakdown channel, the laboratory installation intended for destruction of sub-standard RCPs was developed and manufactured.

The technical and economic destruction indices for commercial products of five types manufactured by Tomsk RCP plants were determined (Table 8.3).

The armature framework of products with average cell size of 250–300 mm was made of steel wires and rods from 4 to 16 mm in diameter. The thickness of the products did not exceed 200 mm; concrete of grades 200–300 was used for their manufacture. Experiments were carried out using the IVG with $W_0 \sim 14.7$ kJ, $U_0 =$ 420 kV, $C_{disch} = 0.167$ μF, $L = 26$ μH, $C \sim 0.033$ μF, and the generator providing 1–2 impulses per second. The sequence of technological operations was the following:

(1) The product was placed on the tray of a technological bath under water layer no less than 200 mm thick; the product was arranged so that the armature hindered

Table 8.3 RCP destruction indices

S. No.	Product type	Metal content, kg/m^3	V, m^3	n_{imp}	W_{spec}, kW h/m^3	q, m^3/h
1	PD–6–2 road plate	70	0.67	1200	7.5	0.30
2	FN–32–14 wall slab	25	0.57	990	7.1	0.50
3	PT–63–12 hollow core slab	30	1.41	2100	6.2	0.70
4	LM–28–12 flight of stairs	140	0.57	1540	11.1	0.25
5	P3–1 voided slab	110	0.16	80	2.0	1.20

as little as possible evacuation of destruction products in the process of concrete removal;
(2) The high-voltage electrode was placed on the product;
(3) The generator was switched on, and concrete under the electrode destructed after application of 1–4 impulses;
(4) The electrode was displaced by 5–10 cm in the transverse direction and by 15–20 cm in the longitudinal direction to a new position.

Destruction was visually controlled by electrode lowering.

The specific energy consumed for product destruction, in kW h/m^3, was calculated from the formula

$$W_{spec} = \frac{n_{inp} W_0}{V}, \tag{8.3}$$

where n_{imp} is the number of impulses applied to the product until its complete destruction and V is the product volume.

To determine the productivity q of installations, timing of technological operations was performed. As a result of experiments, the degree of armature cleaning from concrete for products with two-layer reinforcing was (90–95)% and higher, and for products with single-layered reinforcing, the armature was completely cleaned from concrete. Moreover, the reinforcing framework, including wire armature that required insignificant finalization (correction), was suitable for application in a new RCP.

The character of destruction of some products is illustrated by Fig. 8.6. The experimental values of specific energy consumption and of the installation productivity are presented in Table 8.3. According to these data, the energy consumed for RCP destruction is primarily determined by the degree of saturation of the product with the armature. Thus, for products with single-layer reinforcing, $W_{spec} \approx 2$ kW h/m^3, whereas for products with two-layer reinforcing, these values are 3–5 times higher and change from 6 to 11 kW h/m^3. From the chosen products, the flights of stairs possessing a high degree of reinforcing require the highest specific energy consumption for destruction.

To estimate the degree of concrete destruction, the grain size analysis of destruction products of some types was performed (Table 8.4). As follows from Table 8.4, destructed concrete contained the following fragments: (5–7)% of fraction with sizes (0–5) mm, (45–50)% of fraction with sizes (5–40) mm, and (43–50)% of fraction with sizes exceeding 40 mm (The yield of initial rubble is indicated in brackets.).

The maximum fragment size did not exceed 100–130 mm, and the fragment content was (15–20)%. It should be noted that the initial fillers (sand or rubble) are partially separated from the cement stone already in this stage of RCP recycling. For example, more than half weight of the crushed concrete fraction with sizes (10–20) mm was formed by initial rubble. An acceptable level of cost of specific energy consumption that is approximately 10 times less in comparison with that of mechanical destruction was reached. However, the productivity of the installation was such that constructive and technological measures were required to increase it.

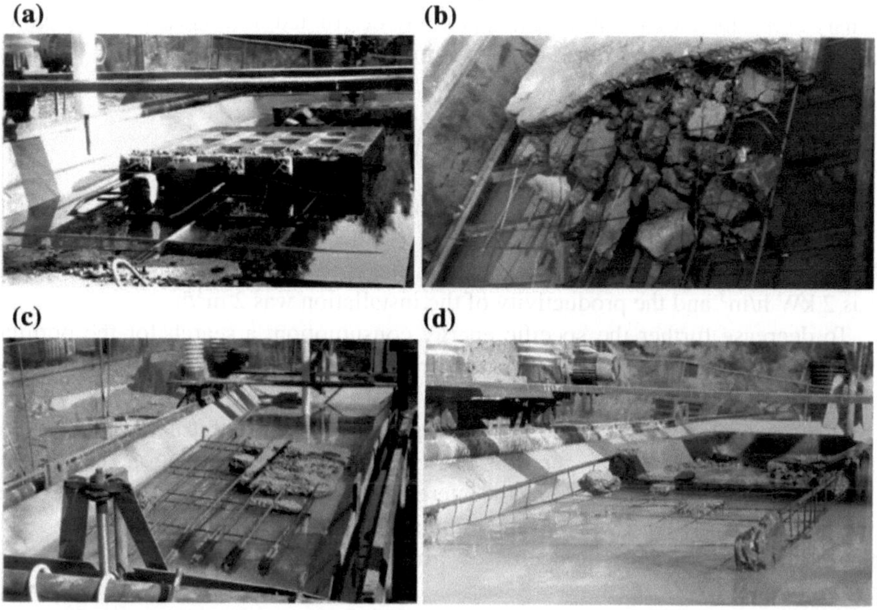

Fig. 8.6 Reinforced concrete products in the stage of armature cleaning from concrete: **a** PD–6–2 road plate, **b** PT–63–12 hollow core slab, **c** LM–28–12 flight of stairs, and **d** P3–1 voided slab

Table 8.4 Grain size structure of destruction products of some types

Product type	Content of fractions (mm), %						
	0–5	5–10	10–20	20–40	40–70	70–100	>100
PD–6–2 road plate	4.9	9.7	21.0 (11.3)	17.0 (2.8)	11.9	17.9	17.6
PT–63–12 hollow core slab	7.9	10.0	21.5 (11.5)	15.9 (2.0)	10.8	17.5	16.4

Results of these tests allowed methods of modernization of ED installations for RCP destruction to be recommended among which the most expedient are:

1. Application of multielectrode systems that will significantly decrease the time of electrode repositioning.
2. Application of automatic systems of mutual displacement of the electrodes (for example, of the electrode and the armature) providing an optimal arrangement of the electrodes and the armature before the application of another voltage impulse.
3. Use of optimal methods and regimes of destruction of a concrete RCP type.

Examples of their implementation in semi-industrial installations are given in Appendix A. For example, to destruct products with single-layered reinforcing, it is expedient that the high-voltage electrode connected to the generator moved continuously along the surface of the product being destructed. In this case, the speed of electrode motion is determined by the formula

$$U = n_{sec}\sqrt{\frac{W_0}{\omega S}},$$

where n_{sec} is the number of impulses per second and S is the thickness of the product.

In experiments on ED destruction of P3–1 voided slabs with $V = 0.18$ m/s, $W_0 = 10$ kJ, $n_{sec} = 1.5$ impulses/s, and $S = 0.1$ m, the specific energy consumption W_{spec} was 2 kW h/m^3 and the productivity of the installation was 2 m^3/h.

To decrease further the specific energy consumption, a search for the optimal parameters of the energy package was required considering the action of each impulse on the destruction of solid materials.

In the considered installation, high energy consumed for destruction was due to non-optimal regime of IVG operation. In the process of RCP destruction, the destruction conditions changed (including the interelectrode distance changing in the process of material destruction), whereas the impulse parameters remained unchanged. The electric energy consumption decreased when using the installation constructed as shown in Fig. 8.7.

The installation comprises IVG 1 whose impulses initiate breakdown in the material. During the breakdown of the material, ICG 5 in which the main portion of energy required for RCP destruction has been stored is switched on; after cleavage of concrete, electrode 3 is lowered down to a certain depth. As the interelectrode distance changed, voltage regulator 6 corrected the IVG voltage providing its operation in such energy liberation regime in which the complex index reflecting the

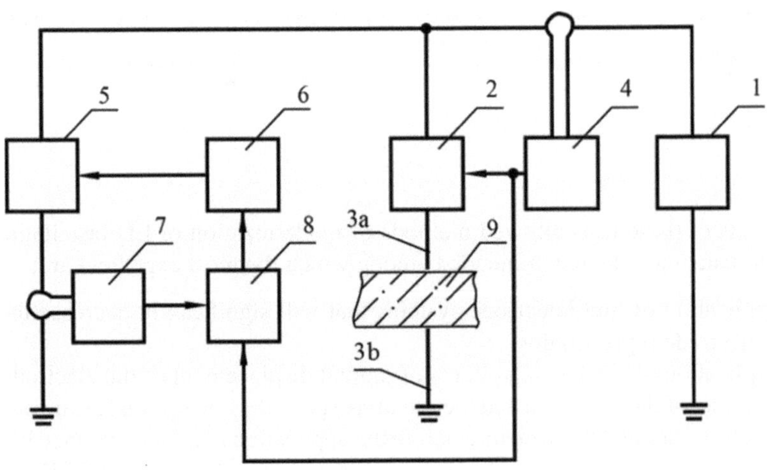

Fig. 8.7 Block diagram of the installation for ED destruction of sub-standard RCP: IVG 1, moving actuator 2, electrodes 3a and 3b, unit for selection and control of output signal 4, ICG 5, voltage regulator 6, current gauge 7, comparison unit 8, and destructible product 9

effect of material properties and discharge circuit parameters on energy deposition remains constant and equal to 0.5. In this regime, the efficiency of the discharge circuit increased by a factor of 1.5.

The efficiency of impulse energy used for destruction can be increased further by a factor of 1.6 by regulation of the generator parameters and of energy transmission from the generator to the load in the matched regime. The impulse source parameters are regulated and matched with the load by a comparison of the wave source impedance $Z_{wi} = \sqrt{\frac{L}{C_{tens}}}$ with the breakdown channel resistance $R_{ch} = \frac{U_{ch}}{i}$, where U_{ch} and i are the voltage drop across the spark channel and the current flowing through the channel.

The above-considered technical solutions and methods of RCP destruction demonstrated that there exist ample opportunities for modernization of the ED installations on recycling of sub-standard RCPs.

8.4 Design Principles and Basic Elements of Installations Intended for Destruction of Reinforced Concrete Products

Results of testing of the laboratory installation for RCP destruction have allowed a semi-industrial installation to be developed to continue investigations on destruction of concrete and separation of armature from concrete to choose optimal regimes of operation under *field* conditions and components and designs of the installation as a whole. The installation operated at the experimental site of the Scientific Research Institute of High Voltage (HVRI), and in the modernized form in Armenia (after the devastating earthquake in 1988; see Fig. A.3 in Appendix A).

8.4.1 Composition and Parameters of the Main Installation Components

Installation specifications:

Energy consumption no more than 70 kW
Specific energy consumption 4–10 kW h/m^3
Productivity 3–5 m^3/h
Degree of cleaning of the reinforcing framework no less than 90%
Fractions of granulated concrete 0–200 mm
Installation mass ~40 ton.

Main assembly units of the installation are:

(1) Distribution box
(2) Step-up rectifier
(3) Impulse voltage generator
(4) High-voltage bus
(5) Technological bath
(6) Electrode system
(7) Operator cabin
(8) Control panel
(9) Hydraulic station
(10) Compressor station
(11) Pumping station
(12) Fence.

Power supply network parameters:

- Voltage 220/380 V
- Frequency 50 ± 0.2 Hz
- Three-phase circuit with dead-ground neutral.

Parameters of the destructible product:

- Length no more than 6000 mm
- Width no more than 1500 mm
- Thickness no more than 250 mm
- Mass no more than 10 ton
- One and two-layer reinforcing.

The installation allowed sub-standard concrete and reinforced concrete products to be destructed, armature to be separated, and granulated concrete partially ready for recycling to be prepared.

The electrode system that transmits impulses from the IVG through high-voltage electrodes to the destructible RCP is the actuator of the installation. The electrode system contains the cartridge (or "comb") of 9–11 electrodes placed on a cart moving along the destructible product using an electric drive, Fig. 8.8.

Manipulation with electrodes allows the destruction zone to be moved over the entire area of the destructible RCP. In this case, the high-voltage electrode, being within the destruction zone, is exposed to intensive mechanical loads and strong electromagnetic fields generated by the breakdown channel in concrete. Therefore, the efficiency of the destruction process critically depends on the reliability and lifetime of the electrode system and primarily, on its insulation.

From 12 above-listed components of installations for RCP destruction, the most important are the IVG and the cartridge of high-voltage electrodes (actuator) together with the control system and its position relative to the destructible RCP.

Fig. 8.8 Electrode system

8.4.2 High-Voltage Impulse Generator

By the start of works on electrodischarge RCP destruction as well as works on other EDT applications, almost unique source of voltage impulses with amplitude of several hundreds of kilovolts and impulse fronts from tenths of a microsecond and half-decay period from units of microseconds was the Marx multistage impulse voltage generator (Marx IVG). To destruct the RCP using the ED technologies, the IVG should provide the formation of breakdown channel in the thickness of concrete and liberation in the channel of the amount of energy sufficient for electric explosion.

Many journal publications, copyright certificates, and patents as well as some monographs (for example, see Malushevsky [3]) are devoted to investigations of various IVG schemes and methods of regulation of its parameters and designs. Here, as well as in the previous Sections of the monograph, only key parameters of concrete IVGs used in this or that technological installations are indicated.

For ED destruction of RCPs the IVG with voltage of approximately 400 kV and energy of about 20 kJ is required. In most installations developed by us, the IVGs were created with dividing throttles based on the bilateral scheme of charging of industrial capacitors to voltages of the order of 50 kV; the air dischargers placed into a glass-epoxy tube with air flushing were used as spark switchers. The external view of IVGs for platform and ladder configurations used in installations for RCP destruction are shown in Figs. 8.9 and 8.10.

In stationary installations, the IVG was placed in a boxing (wagon) and operated either in the self-triggering regime or in the regime with external triggering. In the second case, the spark switches of the first two steps were triggered through circuit-separating transformers.

In particular, in one of the first full-scale installations developed for ED destruction of RCP, the IVG had the following characteristics:

- Maximum charging voltage ±50 kV
- Voltage impulse amplitude 200–600 kV
- Capacitances 0.083 and 0.167 μF
- Discharge circuit inductance 26 μH
- Maximum deposited energy 30 kJ
- Bilateral scheme of charging capacitive storage devices.

Owing to the limited velocity of mechanical motion of the electrode system, the impulse repetition frequency typically did not exceed 1–3 impulses/s.

Fig. 8.9 IVG platform configuration

Fig. 8.10 IVG ladder configuration

8.4.3 Electrode System (Actuator)

8.4.3.1 Requirements for the Electrode System

The actuator of the installation is the electrode system that transmits high-voltage impulses from the IVG through high-voltage electrodes to destructible RCPs. Its design should meet the following requirements (without which the ED technology of RCP destruction and recycling will be obviously noncompetitive with traditional technologies):

- To provide high efficiency and low specific energy consumption
- To possess long lifetime and reliability of operation

- Not to be material- and labor-consuming in manufacturing
- To be convenient for control and easily changeable.

Experiments demonstrated that the minimum specific energy consumption for armature cleaning are obtained when the energy deposited into each breakdown does not exceed the minimum energy necessary for concrete destruction in one cell of the volume armature framework. Usually 2–5 discharges are excited for each electrode position depending on the product thickness and location of the electrode relative to the armature. For impulse repetition frequency of 0.5–2 Hz, the position of the electrode must be changed every 10 s. The drive and its control system should provide mobile scheme of changing the position of the electrode relative to the destructible product, so that the time spent for this operation was an insignificant part of the total time of armature cleaning.

The simplest variant—the one-electrode system, Fig. 8.11—does not meet most of these requirements.

To move the electrode to a new position, it must be moved in three (vertical and two horizontal) directions. The multielectrode system allows a new cycle of impulse

Fig. 8.11 One-electrode system

action on the RCP to be performed before changing the electrode position. A comb of electrodes that encompasses the full width of the destructible product seems to be effective. This allows the product to be treated row by row without horizontal motion of electrodes during treatment of each row. For flat RCP with widths of 1.5 and 3.2 m, 9–15 and 20–30 electrodes are required, Fig. 8.8.

In the installation for RCP destruction, the electrode system configuration is simpler than that in the drilling implement; however, because of a large number of electrodes each representing an independent high-voltage element with insulation, the probability of failure of the entire actuator (the electrode cartridge) increases.

As well as the drilling implement, the electrode system of the installation for RCP destruction operated under very severe conditions upon exposure to large dynamic loads, strong electromagnetic fields, hydraulic flows, and products of destruction of the material being treated. The main reason for failure of the electrode system is breaking of insulation of the high-voltage electrode under the action of surface discharges and of the main discharge in the region of contact between the electrical conductor and the insulator provoked by them. The destruction starts in the front-end part of the insulator at the dielectric and metal interface. First traces of discharges originate—tracks that with increasing number of impulses applied to the electrode are gradually transformed into deep cracks, Fig. 8.12. Cases of insulation destruction by a cross-cutting breakdown were not observed.

Figure 8.13 shows the sketch of the high-voltage electrode with the simplest configuration being part of the electrode system of the installation for RCP destruction. The electrode represents steel core 1 in polyethylene insulation 2. The core–electrical conductor 1—is insulated almost along the entire length except for tip 3 in the discharge gap and the opposite end to which the IVG is connected.

On the one hand, the insulation allows the energy losses to be minimized because of the conductivity current flowing through water thereby preventing breakdowns to the technological bath and neighboring electrodes; on the other side, it provokes discharges on the insulation surface arising in the "threefold point."

In analogous electrode systems, the electric parameters of the discharge circuit (voltage and interelectrode gap length) and the constructive characteristics (geometrical sizes of the core–electrical conductor, insulation, their relative positioning, and

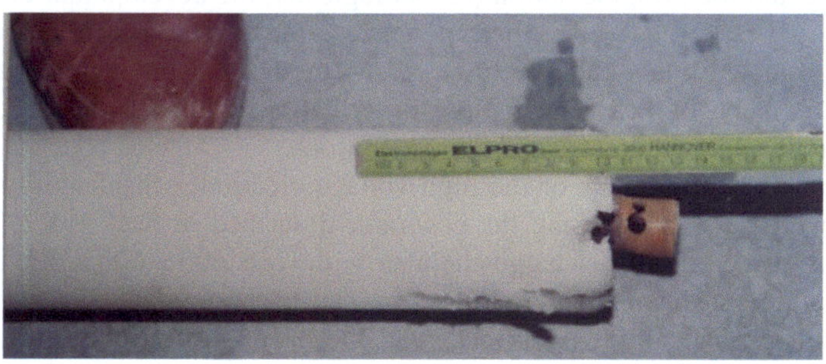

Fig. 8.12 Photograph of the high-voltage electrode after exploitation

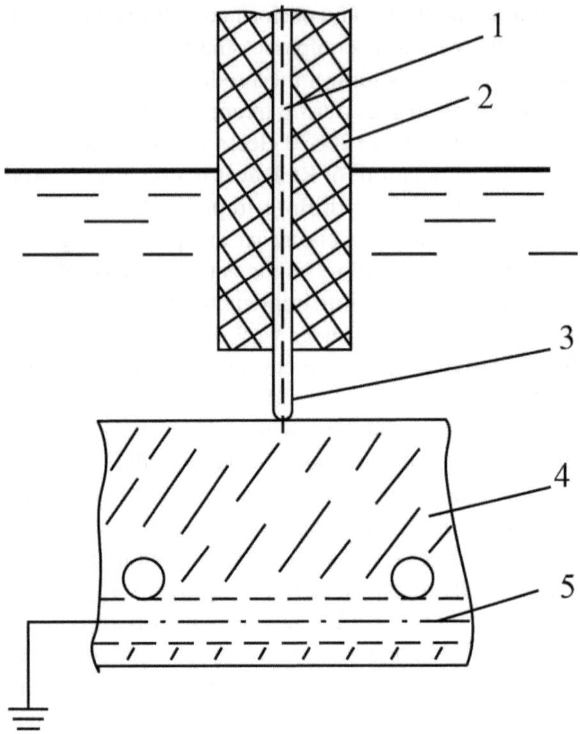

Fig. 8.13 Sketch of the single high-voltage electrode comprising steel core (electrical conductor) 1, polyethylene insulation 2, tip 3, RCP body 4, and grounded armature 5

the area of the bared electrical conductor surface) influence significantly on their reliability and efficiency of their operation.

8.4.3.2 Tests of Electrodes of Various Design

In the initial stages of development of the technology of ED destruction of RCPs, the electrodes were designed based on the results of testing of installations for ED granulating and rock fragmentation (chronologically, the development of this technology started earlier) as well as the experience in the development and operation of bushings, input terminals, cable cuttings, etc. Methods of adaptation of these designs to the requirements of the ED technology of RCP destruction were tested for various modifications under laboratory and field conditions.

Laboratory investigations of the high-voltage electrodes insulated by different methods were carried out using the generator described in Sect. 8.4.2. The electrodes were placed in the discharge chamber with sizes 600 × 600 × 600 mm made of sheet stainless steel with thickness of 3 mm. The discharge chamber had two observation windows made of Plexiglas and located in the opposite walls of the chamber. Experiments were carried out with the core-plane electrode system formed by the tubular electrode with the tip fastened to the chamber cover and the metal bottom of the chamber. Tap water with specific resistance $\rho = 1.86 \times 10^3\ \Omega$ cm was the working medium.

Designs of the tested electrodes are shown in Fig. 8.14. As a core–electrical conductor, water-gas St-3 grade steel pipes with external diameter of 27 mm were used.

The electrical conductors were baked into high-pressure polyethylene insulation which then was treated on a lathe to the diameter $D = 80$ mm.

The electrode length did not exceed 1000 mm. The electrical conductors had replaceable stainless steel tips. The working rounded tip ends had a radius of 11 mm.

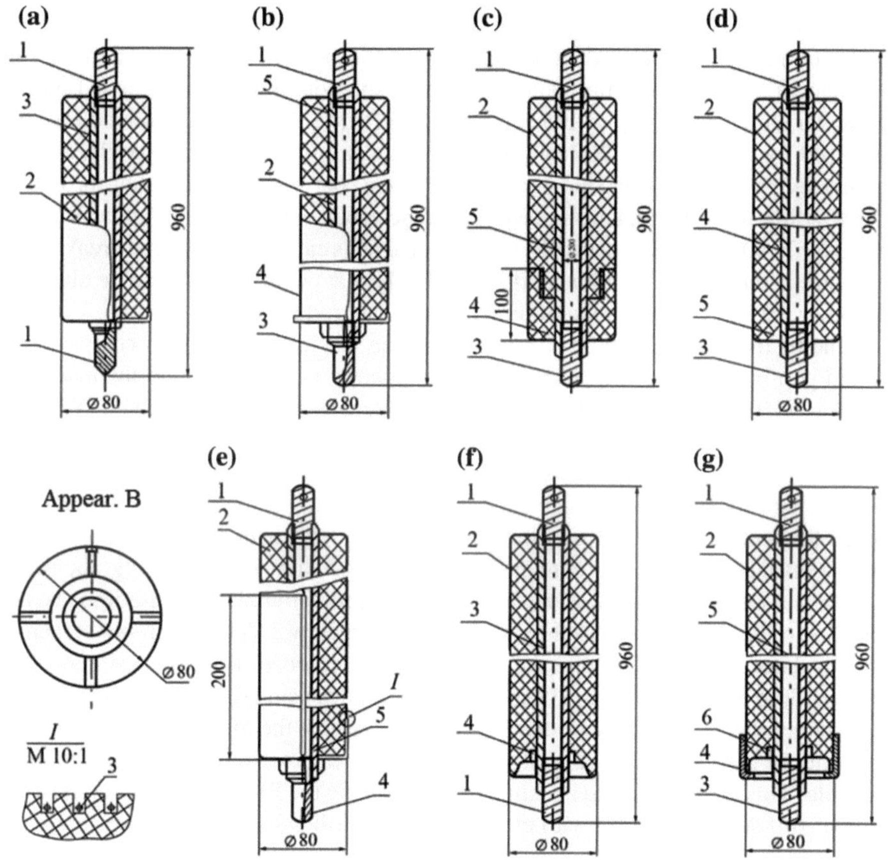

Fig. 8.14 Sketches of the developed and tested electrodes intended for ED destruction of RCP comprising **a** tip, core, and flat end face of the insulator, **b** electrode with a flange, **c** electrode with replaceable insulation tip, **d** electrode with inverted insulator cone, **e** electrode with nichrome wire winding, **f** electrode with radial end face gap, and **g** electrode with radial slotted gap

High-voltage impulses with amplitude $U_0 = 270$ kV and energy $W_0 = 1.5$ kJ were applied to the electrode of the IVG with discharge capacity of 0.042 μF and discharge circuit inductance of 11 μH. The interelectrode gap was 50 mm.

During experiments, the voltage U between the electrodes, the discharge current I, and the discharge time were recorded. After application of a series of high-voltage impulses (from 10 to 100 impulses in a series) to the electrode, the state of the insulator and electrode as a whole was visually analyzed. The experiment was stopped after occurrence of the first traces of destruction at the insulator end face. The time to the discharge, the voltage, and the current were determined after processing of no less than 15 impulse waveforms for each electrode design. Both model and working electrodes of actual technological installations were tested.

As is well known, one of the methods of prevention of discharges over the surface of insulation of high-voltage installation is obtaining of compulsory favorable field distribution using semiconductor coatings. In the present study, we attempted to implement this idea by bifilar winding on the insulation of nichrome wire 0.1 mm in diameter with resistivity $\rho = 1.05\ \Omega\ mm^2/m$, Fig. 8.14e. The number of winding coils was 100 for the insulator 200 mm long. The lower end of the wire was electrically connected to the electrical conductor; the resistance between the beginning and end of the winding was 2.8 kΩ. During testing, the electrode withstood without destruction traces only 40 high voltage impulses. Continued tests caused melting and breakage of the wire at the insulator end face. It seems likely that positive result can be obtained by other methods of field regulation along the electrode insulation.

One of the simple methods of increasing the electrode lifetime is creation of a liquid gap in the form of a funnel or a hemispheric insulator around the electrical conductor. Testing of such electrodes actually demonstrated their high strength; no traces of discharge and destruction arouse on the insulator end face.

In the process of destruction of a reinforced concrete product, when the high-voltage electrode was submerged deep into the product, the insulator could touch the upper layer of the grounded armature. In these cases, the main discharge can develop on the insulator surface and damage it. In the design shown in Fig. 8.14g, the metal screen shaped as a beaker was used for insulator protection. During lifetime tests of the electrode, the crack caused by sliding discharges was detected in the screen at the insulator end face. Despite the short lifetime recorded in this experiment, it is expedient to continue investigation of such electrodes using screens of other shapes.

Among the developed and tested electrode designs, the most reliable electrodes with liquid gap in the insulator at the electrical conductor have proved to be most reliable. This idea was tested in actual practice (using the ED installation for destruction of sub-standard RCPs) with electrodes of the following designs, Figs. 8.15 and 8.16:

- Radially slotted gap at the electrical conductor
- Cone-shaped insulator end face (water interlayer shaped as a funnel)
- Tubular screen and water layer.

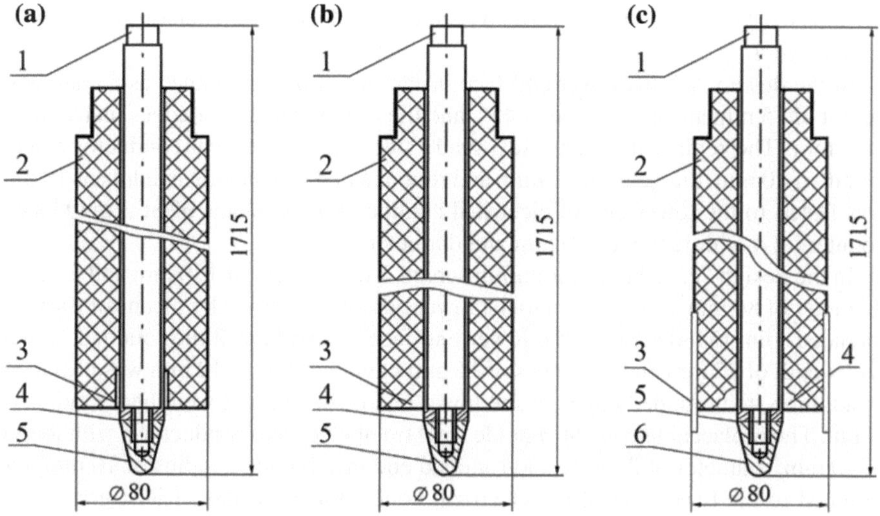

Fig. 8.15 Design of insulators tested under semi-industrial conditions: **a** with a radially slotted gap at the electrical conductor, **b** with a cone insulator end face (a water layer shaped as a funnel), and **c** with a tubular screen and a water layer

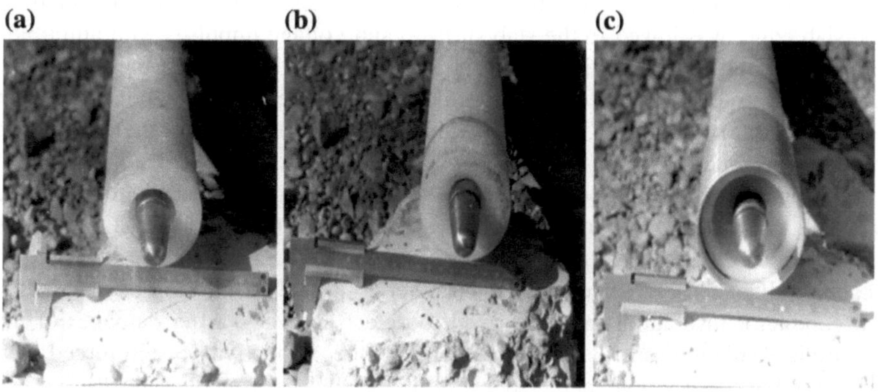

Fig. 8.16 Photographs of the real electrodes schematically shown in Fig. 8.15

The main elements of the electrodes were identical: core–St 3 grade electrical conductor 1 1650 mm long and 27 mm in diameter placed inside of high-pressure polyethylene insulator 2 1590 mm long and 80 mm in diameter.

In design "*a*" (Figs. 8.15 and 8.16), annual groove was provided forming around core–electrical conductor 1 radially slotted water layer 3 30 mm long and 5 mm wide at the lower end face of insulator 2. Replaceable rounded St-3 grade steel tip of electrical conductor 5 with radius of 10 mm, length of 40 mm, and diameter of

22 mm was fastened to the lower end face of the electrical conductor by means of counternut 4.

In the design "*b*," the lower end face of the insulator was shaped as a cone with height of 15 mm and apex angle of 45°, and the rounded insulator edges had a radius of 5 mm. The replaceable cone St-3 grade steel tip of electrical conductor 5 with length of 40 mm, diameter of 22 mm, and rounded end face having a radius of 10 mm was fasted to the lower end of electrical conductor 1 by means of St-3 grade steel counternut 4 with length of 10 mm and diameter of 27 mm.

In the design "*c*," tubular duralumin screen 3 with length of 150 mm and external diameter of 86 mm was fastened to the lower part of insulator 2 by means of a screw-threaded joint (M80 × 2). At the lower end face of insulator 2 the rounded groove with depth of 25 mm and thickness of 16 mm having radius of 12 mm was provided. In addition, the rounded edges of the lower end of insulator 2 were had a radius of 3 mm. The replaceable cone St-3 grade steel tip of electrical conductor 6 with length of 40 mm, diameter of 22 mm, and rounded end face having a radius of 10 mm was fastened to the lower end of the electrical conductor by means of 3-St grade steel counternut 5 with length of 10 mm and diameter of 27 mm.

The tested electrodes were placed in the middle of the cartridge ("comb"). Sub-standard PT–63–15 slabs with two-layer reinforcing were destructed. They were loaded into the technological bath filled with technical water with specific resistance of 3×10^3 Ω cm. The tested electrode was placed at a distance of about 150 mm from the slab edge, was lowed to the slab surface, and voltage impulses with amplitude of 350 kV and energy of 10.2 kJ were applied from the IVG. In the process of concrete destruction, the electrode moved down. After destruction of concrete under the electrode, it moved up, was displaced to a new position, and the destruction process was repeated.

During testing, the number of impulses withstood by the electrode before damage was determined. High-voltage impulses were applied with period of 1–2 impulses/s in series of 100 impulses in the beginning of the experiment and 50 impulses in second half of the experiment. After each series, visual monitoring of the electrical conductor insulation was performed to detect damages. After damages of the electrode insulation (cracks, tracks, and so on), the productivity of reinforced concrete product destruction sharply decreased, and the experiment stopped. The last series of impulses, under application of which the electrode failure was observed, was disregarded in lifetime calculation. Three electrodes of each design were tested together with the electrode of conventional design with flat insulator end face, Fig. 8.13. Results of electrode testing on the semi-industrial installation are presented in Table 8.5.

The character of electrode insulation destruction was the following. Discharge traces in the form of several cracks appeared on the electrode with flat insulator end face after application of 600–700 impulses.

From 2 to 3 carbonized discharge traces with depth of 1.5 mm and width up to 2 mm appeared on the insulator end faces of electrodes with radially slotted gap.

Electrodes with cone insulator end face withstood up to 900 impulses.

Table 8.5 Electrode lifetimes

	Electrode design and number of impulses before failure (average lifetime)			
Electrode number	Insulator with flat end face	Insulator with radially slotted gap	Insulator with cone end face	Insulator with tubular screen
1	650	800	900	1250
2	700	750	850	1300
3	700	850	950	1200
Average lifetime	683	800	900	1250

Thus, testing of the developed electrodes demonstrated an increase in the lifetime by a factor of 1.2–1.3 in comparison with the electrode with flat insulator end face. The electrode with tubular screen was the most effective one since it allowed the lifetime to be increased by a factor of 1.8.

In the process of further modernization of the electrode design, the electric field near the threefold (polyethylene-water-steel) point was investigated using computer simulation.

8.4.3.3 Electric Field Distribution in the Working Zone of the Electrode and Its Regulation

Computer simulation was performed using the software package ELCUT—the integrated dialogue program package that allowed flat and axisymmetric problems to be solved including calculation of electrostatic fields, fields of direct and alternating currents, and so on.

The input data for simulation were taken proceeding from the actual sizes of the design: 45 mm interelectrode gap (the thickness of concrete to the armature in the base of the reinforced concrete sleepers), core–core electrode system, and steel high-voltage electrode Ø 27 mm in polyethylene insulation Ø 80 mm. The working voltage was 450 kV.

The program package ELCUT allowed the results of simulation to be displayed as an electric field pattern (Figs. 8.17 and 8.18) and E to be measured at any point of the simulation region.

One of the methods of decreasing the electric field strength at the threefold point is to increase the length h of the uninsulated part of the electrode (Fig. 8.19).

The value of E at the threefold point can be decreased using the constructive approach—sharpening of the insulator end face for the cone with the apex angle $\alpha =$ 45°–80°, Fig. 8.20.

One more important factor of decreasing the field strength at the threefold point is the geometry of uninsulated part of the electrode. Figure 8.21 shows the pattern of the field for the electrode with cone-shaped tip. From Fig. 8.21 it can be seen

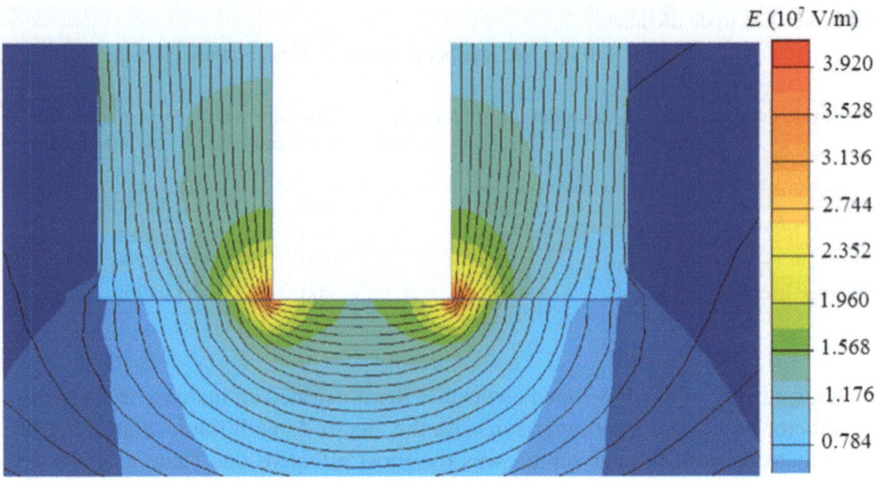

Fig. 8.17 Electric field strength at the threefold point near the high-voltage electrode

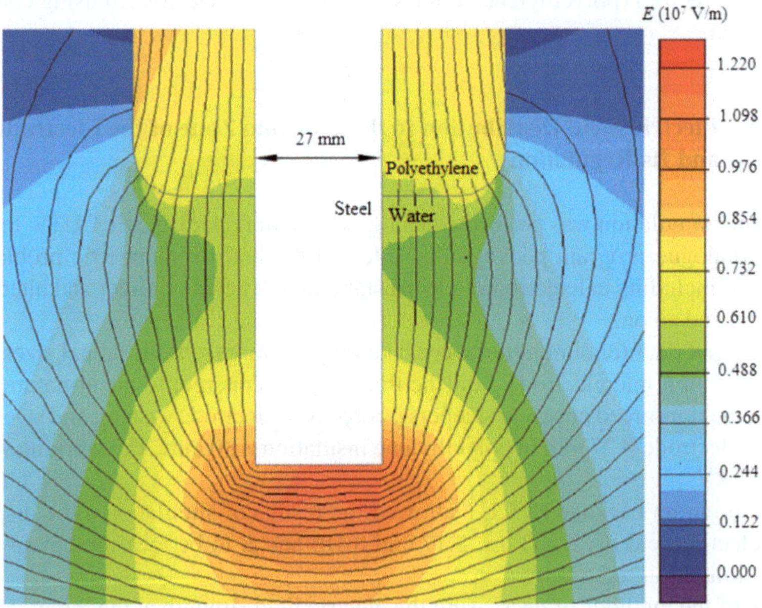

Fig. 8.18 Electric field pattern for the electrode with length of non-insulated part of the electrode $h = 58$ mm

that E at the threefold point is significantly lower than for the previous high-voltage electrode designs.

Results of simulation and testing have allowed us to formulate the following recommendations for the electrode design:

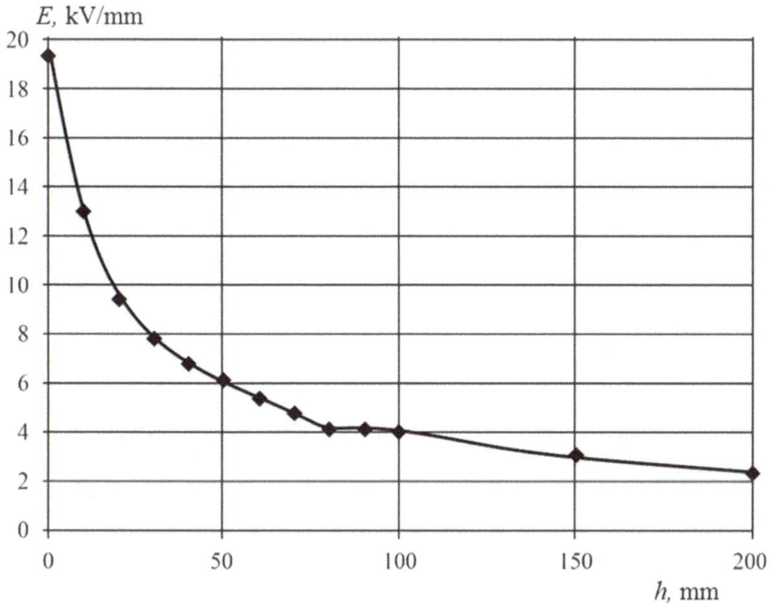

Fig. 8.19 Dependence of the electric field strength at the threefold point on *h*

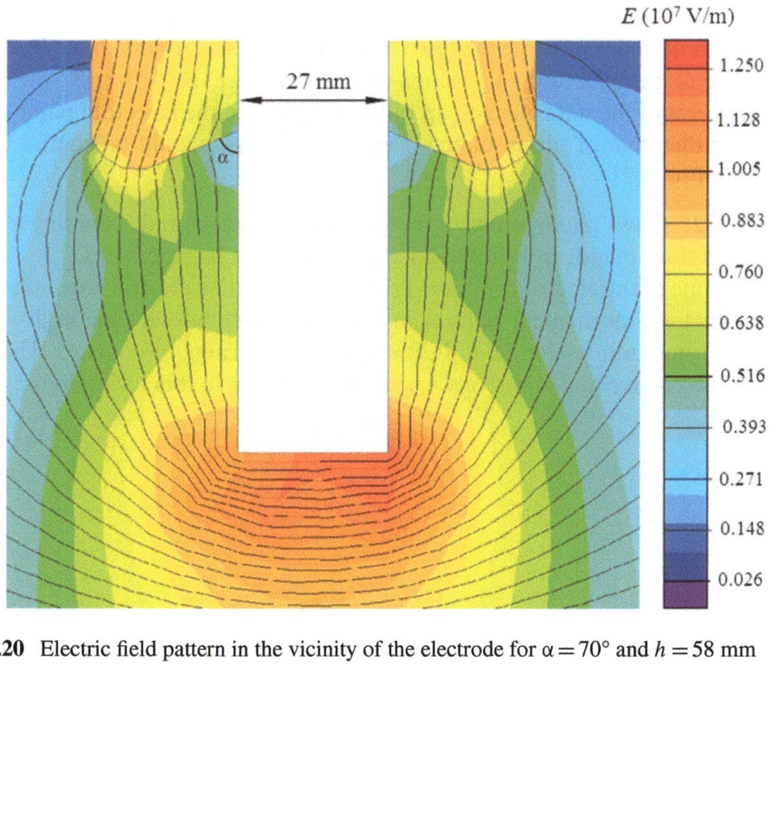

Fig. 8.20 Electric field pattern in the vicinity of the electrode for $\alpha = 70°$ and $h = 58$ mm

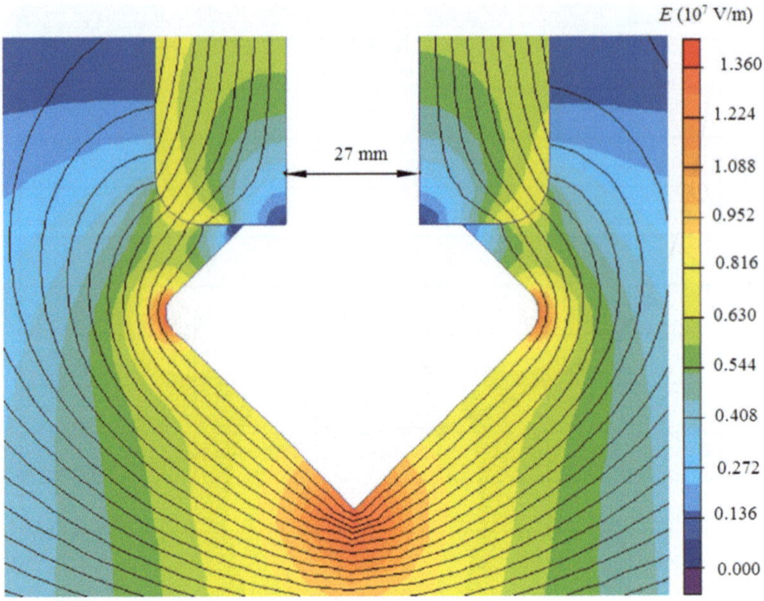

Fig. 8.21 Electric field pattern for the electrode with cone-shaped tip for $h = 58$ mm

- The length of uninsulated part of the electrode should be no less than 60 mm;
- The angle at the apex of the insulation sharpening cone should be 45°–80°;
- The diameter of the uninsulated part of the electrode at the threefold point should be larger than that of the insulated part, and its shape should be similar to that shown in Fig. 8.21.

Figure 8.22 shows photographs of the electrode manufactured according to these recommendations that demonstrated good performance characteristics.

The lifetime of the electrode system is also influenced by electric erosion of electrical conductors, mechanical failures in fastening assemblies of the high-voltage electrode during motion along the RCP surface, and burning of contacts. This suggests that the erosive and abrasive wear of the electrode in this technology is not a big problem.

Fig. 8.22 Electrode with optimized geometry in the working position

8.4.4 Control of the Position of the Electrode System Relative to the Product to be Destructed

Considering the presence of metal armature in RCPs, it is expedient to use it as the second—earthed-electrode. In the initial state the electrode system is close to the "core-plane" system. In the process of concrete destruction, the interelectrode gap gradually decreases, and the electrode system by the character of electric field becomes similar to the "core–opposite core" or "core–core" system placed on one surface. In practice of recycling of rejected RCPs, the first two types of electrode systems are most often encountered; the letter is possible when RCPs with multilayered densely located armature are destructed.

The system of control over the position of high-voltage electrodes should provide their motion in three mutually perpendicular directions, and the scheme of voltage application to them should provide the distribution of the electrode potential so that the breakdown was initiated at the required point. Two approaches are possible here:

- All electrodes are located above the product being destructed and, by means of the special device—the distributor—the voltage impulses are successively applied to each electrode. The system is bulky and complicated and so far we failed to embody it "in metal";
- Only the electrode that initiates the breakdown of concrete in the forthcoming destruction cycle is placed on the RCP surface. During its motion (usually vertical), it is connected to the high-voltage bus. The potential can simultaneously be

applied to all electrodes, but in this case, all electrodes, except one, are arranged in positions, in which only one electrode in contact with the destructible RCP operates.

For multielectrode systems of cartridge (or "comb") type, the main motion is vertical; the longitudinal motion of the entire comb is performed periodically, after destruction of the product in a row; the transverse motion is required if the product width is greater than the comb length. It is expedient to perform transverse motion within the step of the armature network to change the position of the "comb" in case of wrong orientation of electrode motion relative to the network.

Manipulations with electrodes relative to the destructible RCPs can be performed using an electric-, pneumo-, or hydrodrive. The *pneumodrive* has high operating speed, is not sensitive to electrical interferences, and the energy carrier (compressed air) is available practically in each enterprise of building industry. The *electric drive* is simpler for application, more compact, but is sensitive to interferences and requires insulation. It is difficult to replace the *hydrodrive* to eliminate the product to control over the results of treatment and tray unloading.

It is most expedient to move the electrodes in the vertical direction using the pneumodrive. In collaboration with the Design and Technology Bureau Scientific Research Institute of Field Research Studies of Reinforced Concrete (Moscow), we have developed and tested the drive in which the electrodes were used as pistons of cylinders. This allowed us to produce sufficiently compact electrode system intended for operation as a part of a mobile installation (on a railway platform). However, during electric explosion, shock loads through the electrodes were transferred to elements of the power cylinder thereby limiting its lifetime.

In stationary installations the standard power cylinder, whose rod is connected to the electrode through an elastic dielectric insert made of plasticized Capron threads, was used. The insert unloads elements of the pneumodrive from shock loads and also prevents possible overflows of charges through the "piston-cylinder" interface. The transverse cartridge motion (by 150 mm) was restricted by the standard pneumo-cylinder or by the electric drive through a "screw-nut" pair using the belt drive of the engine shaft. (The belt withstood without breakdown and flashover the voltages up to 400 kV.) The transverse stroke was 1500 mm that allowed products with a width of 3.2 m to be destructed using an 11-electrode cartridge.

It is most expedient to determine the longitudinal stroke of the electrode system relative to the motionless RCP or of the cart with the product relative to the motionless electrode system using standard technical solutions applied to the electric drive. The ED technology allows the process to be automated, the energy consumption to be significantly decreased, and the productivity to be increased. With application to the operation of armature cleaning, the automatic control should provide high relative number of "useful" discharges—initiated in the thickness of concrete—in comparison with the relative number of "useless" breakdowns—over the surface between the high-voltage electrode and the bared armature, between the high-voltage electrode and the technological bath, and so on.

The optimal point of high-voltage electrode arrangement on the product surface for a periodic structure of the reinforcing mesh is such when the projection of the point of arrangement onto the plane of the nearest mesh is the focus equidistant from the armature elements. This corresponds to the maximum probability of concrete breakdown and the greatest length of the breakdown channel in the concrete thickness.

To construct an automatic control system, it is necessary to choose a suitable parameter that yields reliable data on localization of the breakdown channel in the thickness of concrete, on its surface, or in water as well as of the length of the breakdown channel. In practice of monitoring of the high-voltage energy equipment, such parameters as gas liberation (its intensity and composition of gases), electromagnetic and acoustic emission, and so on are used. The application of these parameters in the ED technology of RCP destruction is impeded by the fact that they require complicated identifying equipment, are applicable only for a narrow range of variations of external conditions and for periodic graduation when the conditions change.

Another parameter—the difference between the breakdown channel resistivities (per unit length) for solid dielectrics (including concrete) and liquids and in the surface flashover—turned over to be more suitable. In the first case, it was approximately by a factor of 1.5 higher. This allowed cases of "useful" and "useless" breakdowns to be distinguished by the discharge current amplitude and waveform and the parameters of installations and the regimes of their operation providing the maximal efficiency of the discharge circuit (up to 0.5–0.7) to be chosen.

8.4.5 Preparatory Technological Operations

The first operation in the technology of RCP destruction is placing of the product in the technological bath filled with water with definite orientation of the armature relative to the electrode system. In the process of finishing of the design and testing of installations, the product was placed on the tray (for example, shaped as a trough) to the bottom of which the transverse stiffeners serving simultaneously as supports for the destructible product were welded. The inclined tray sides prevented spreading of fragments in the process of concrete destruction. Slits in the tray sides provided water discharge when the tray rises over the bath. The space between stiffeners was calculated to collect concrete fragments in the process of RCP destruction.

For the motionless bath, it is expedient to arrange guiding rails along which the electrode system can move. If the electrode system is motionless, the bath is mounted on the cart. Arrangement of the electrode system on a bridge crane moving along the bath seems to be an efficient (yet unapproved) technical solution.

To control periodically the efficiency of RCP destruction and to make loading and unloading operations easier, it is expedient to establish the tray on the power cylinders (using the hydraulic drive).

By means of the drive of transverse motion, the electrode system is placed so that each electrode after longitudinal displacement was in the middle between two longitudinal bars of the metal armature.

The minimal energy consumption corresponds to the case of bypassing the product by a spiral from the periphery toward the center; in this case, breaking on the free surfaces is used most fully. For the "comb" type electrode system, manipulations for such bypassing are rather complicated; it is more expedient to destruct the product row by row from one free surface to the opposite one.

8.5 Performance Indicators Semi-industrial Installations

The degree of cleaning of the armature for products with two-layer reinforcing is 90–95%; it is 100% for products with single-layer reinforcing. As already indicated above, the armature network after insignificant finishing is suitable for recycling in RCPs.

Figure 8.6 shows fragments of the destructed and partially cut products, and Table 8.3 presents the main quantitative results of testing. From 3 to 5 industrial products manufactured by two Tomsk enterprises of panel housing construction were destructed. These destruction characteristics for several products of one type yield the upper estimation, because they were obtained without optimization of energy deposition. The energy per impulse was 19 kJ, $U_0 = 420$ kV, the impulse repetition period was 1 impulses/s, the number of impulses per cycle (for one electrode position) changed from 1–3 to 4–6, and the time spent for changing of the electrode position was 25–30 s (depending on the type of product to be destroyed).

The main factor determining the specific energy consumption (and the productivity of the installation) was the degree of product saturation with the armature. Thus, for products with single-layer reinforcing in the form of a network, $W_{spec} =$ 2 kW h/m^3, and for products with volume two-layer reinforcing, $W_{spec} = 8$ kW h/m^3.

To substantiate the efficiency of destruction of products saturated with the armature (columns, crossbars, and so on), additional investigations are required. Here the penetration depth for the electrodes into the thickness of products not always provides the achievement of the expected effect. Softening of concrete is reached by treating of the surface layer; however, its removal from the internal regions faces certain difficulties. An analysis of the experience of Tomsk building industry enterprises demonstrated that the probability of deviation from the standard regime of manufacture in all stages of RCP production is the higher, the lower is the degree of product saturation with the armature. This suggests that the most part of substandard products can be treated with energy consumption in the stages of cleaning of the armature that did not exceed 10 kW h/m^3.

Table 8.4 gives the particle size distribution of destruction products of PD–6–2 road plates and PT–63–15 hollow core slabs. The initial rubble yield is indicated in brackets. From the table it follows that initial fillers already in the stage of armature cleaning are partially separated from the cement stone. For example, more than half (by weight) of the 10–20 mm fraction of crushed concrete was formed by initial rubble.

The decrease of the energy per impulse down to $W_0 \sim 14.7$ kJ in the process of testing of PD–6–2 road plates has led to the increase in the 100–130 mm fraction up to 30–35% (specific energy consumption decreased from 8.6 to 7.5 kW h/m^3, and the productivity decreased from 0.3 to 0.26 m^3/h per one electrode).

In the initial stage of RCP destruction when the high-voltage electrode was on the product surface, the most probable direction of the discharge channel evolution was through the concrete thickness toward the armature, since other grounded elements were located at much larger distances from the high-voltage electrode. However, in the process of concrete destruction the elements of the armature toward which the breakdown channel will be oriented either in the water thickness or on the concrete surface will be bared. In this case, the efficiency of the process sharply decreases, since in the acoustic approximation only about 35% of the wave energy generated in water is transferred to concrete due to refraction at the interface between two media. In this regard, the parameters of the high-voltage impulses acting on the electrodes and also the method of manipulation with them should be chosen so that to provide the maximal probability of breakdown in the concrete thickness and the maximum possible length of the spark channel. In combination with the optimal parameters of the energy package in the final breakdown stage—the stage of energy deposition—this allowed the specific energy consumed for destruction to be minimized.

As follows from Chaps. 3 and 4, at $t \approx 10^{-7}$ s the breakdown gradients of technical water for impulses of positive polarity were by a factor of 1.5–2 greater than those of concrete. With such voltage impulse duration, the breakdown in the system "high-voltage core–grounded armature" will be observed in the concrete thickness rather than in water or on the concrete surface for equal distances on the surface and in the concrete.

Our investigations demonstrated the expediency of application of high-voltage sources with different amounts of the deposited (and liberated) energy for destruction of RCPs with very different parameters. Thus, for example, to destruct pipes, trays, and simplest designs with single-layer reinforcing, the energy equal to 5 kJ was sufficient; to destruct wall slabs, road plates, and other products of analogous assortment, the energy equal to approximately 20 kJ was required. To destruct beams, crossbars, and other elements of building industry with high degree of reinforcing, still higher amount of energy will be required. The expediency of EDT application for the destruction of such products has not been tested yet. It should be born in mind that the relative amount of such products in the sub-standard products is very small.

References

1. Kuretz VI, Usov AF, Zuckerman VA (2002) Electropulse disintegration of materials. Publishing House of the Kola Scientific Center of the Russian Academy of Sciences, Apatity, 324 p (in Russian)
2. Yutkin LA (1986) Electrohydraulic effect and its application in industry. Publishing house Maschinostroenie, Leningrad, 253 p (in Russian)
3. Malushevsky PP (1983) Fundamentals of discharge-impulse technology. Publishing house Naukova Dumka, Kiev, 272 p (in Russian)

Chapter 9
Fragmentation of the Filler—The Final Stage of Preparation of Reinforced Concrete Products for Utilization

9.1 Choice of the Method of Fragmentation of a Concrete Filler

The choice of an effective method of recycling of pieces of concrete into the cast fractions formed in the process of destruction of sub-standard RCP is important in connection with great volumes of works on reconstruction of buildings and structures.

Recycling waste should be involved into the economic cycle to better meet needs in raw materials of the building industry and to protect the environment from technogenic pollution and damage to the natural landscape. Technological operation of fragmentation and disintegration of solids—rocks, ores, and artificial materials—is widespread in industry. In this case, the ultimate goals of these operations depend on the branch of industry, namely, in preparation of ores for enrichment, the main factor is the selectivity of destruction; in chemical industry, this is the size distribution and purity of the end product; and in building manufacture, in the construction industry this is the shape and condition of surface of grains of definite fractional composition. The above-indicated purposes and the physicomechanical properties of the product being fragmented and sizes of its pieces determine the choice of the most effective and economic fragmentation method.

Classification of numerous methods of fragmentation and disintegration of solids depending on sizes of the product being fragmented, and employed types of energy and physical principles is presented in a number of monographs and scientific articles.

Distinction is made by sizes of the initial and final fragmented products:

- Coarse fragmentation from fraction 300–1500 mm to fraction 100–300 mm,
- Medium fragmentation from 100–300 mm to 40–100 mm,
- Small fragmentation from 40–100 mm to 5–30 mm,
- Fine fragmentation with initial sizes of several millimeters.

V. Y. Ushakov et al., *Electro-discharge Technology for Drilling Wells and Concrete Destruction*, https://doi.org/10.1007/978-3-030-04591-3_9

For ***coarse fragmentation***, the following methods are mainly used:

- Mechanical methods realized in crushing machines, guillotine-type machines, and pneumohammer devices,
- Explosive methods,
- Thermal methods including electrothermal ones,
- Electrohydraulic and electrodischarge methods that are being developed and mastered.

For ***fine fragmentation***, traditional mechanical methods are widely used and radically new methods are being introduced, including

- Jet method of self-fragmentation of particles in collisions of pieces and particles of the material being fragmented,
- Jet-impulse technology of material dispersal under the action of short-term application of high-pressure (0.4–1.0 MPa/m^2) compressed gas impulse and impact on a baffle slub,
- Pneumomechanical method of fragmentation by pressure in which particles are destructed when liquid filling particle cracks boils up,
- Snyder's process (combination of the explosive and jet methods) in which the material exposed to vapor or gas is ejected with a speed of about 300 m/s through accelerating pipes into a large-capacity collecting vessel where it instantly depressurizes and collision occurs of its opposite jets,
- Cavitation method of fragmentation of the material in a liquid by mechanical excitation of hydraulic shock waves or excitation of ultrasonic oscillations.

For medium and small fragmentation, both mechanical and electrophysical methods are used and are being developed, including electrohydraulic and electrodischarge methods used for small fragmentation. As to fragmentation of concrete fraction 70–150 mm down to sizes less than 40 mm, this method is referred to as small and medium fragmentation.

9.1.1 Mechanical Methods of Fragmentation

Devices for fragmentation and disintegration used in concentrating mills are subdivided into five main classes:

- Jaw crushers,
- Cone crushers,
- Roll crushers,
- Hammer crushers and disintegrators,
- Tumbling mills.

An advantage of the mechanical crushing devices is their high efficiency. Thus, Traylor Broth., Inc. (Great Britain) produced a cone crusher with productivity of

2300 t/h. The device is intended for primary crushing of granite down to sizes less than 15–20 mm used then as filler.

The crushing methods and devices used in open-cast mines include crushing machines, pneumatic hammers, punchers, hydraulic presses, and drop hammers (destructing by falling loads) also differ [1]. Pneumatic and hydraulic devices are efficient only for some kinds of works. The method of crushing by falling load possesses low energy efficiency and productivity because of chaotic arrangement of pieces being crushed and low efficiency of energy transfer to a destructible piece when the load falls on its inclined surface.

As a whole, we can state that traditional mechanical methods of destruction of rocks and artificial solid materials (for example, concretes) are practically close to the limit of their technical possibilities. The devices, implementing these methods, differ by complex designs, strong wear of working parts, high energy consumption. For these reason, their application for secondary fragmentation of concrete is problematic.

9.1.2 Electrohydraulic (Electro explosive) Method of Fragmentation

In principle, solid materials can be destructed by ultrasonic oscillations, high-frequency electric and magnetic fields, and low-frequency currents. A number of disadvantages characteristic for these methods (often called electrophysical ones), including high energy consumption for destruction (up to 80 kW h/m^3 for destruction by low-frequency currents), large overall dimensions and weight of installations, and production factors adverse for health of personal prevent their wide application. The most widespread from the electrophysical destruction methods is the electrohydraulic (EH) method based on the destruction effect of pressure waves produced in the discharge channel in water [2]. Impulse current generators with operating voltage from 5 to 50 kV and energy per impulse up to 10^5 J are used as energy sources. During the discharge, the current running through the discharge channel of such energy source being of the order of 10^4 A heats it to a temperature of $(20–40) \times 10^3$ K when the pressure increases up to $(3–10) \times 10^2$ MPa [3].

The main disadvantage of the EH method is extremely high specific energy consumption, especially for fragmentation of strong materials, caused by the presence of the intermediate medium (a liquid) in the process of electric energy conversion into the energy of newly formed surfaces of the destructible solid, Table 9.1.

As follows from Table 9.2, the specific energy consumption for the EH technology exceeds by an order of magnitude and more the energy consumption for the traditional destruction methods.

Nevertheless, the EH technology can be referred to progressive (innovative) one since it possesses two important technological advantages: (a) allows the end product with narrow particle size spectrum to be obtained, and (b) minimizes product

Table 9.1 Energy consumption for fragmentation by the electrohydraulic method

No.	Material	Initial size, mm	Final size, mm	Energy consumption, kW h/t	References
1	Limestone	−31 + 10	0.2	370.0	[4]
2	Metallurgical slag	+20	0.2	2000.0	[4]
3	Gravel	10	0.3	50.0	[5]
4	Quartzite	200	30.0	100.0	[6]
5	Quartzite	10	0.2	66.7	[6]
6	Sherlovogorsk ore	−30 + 0	−2 + 0	117.2	[7]

Table 9.2 Specific energy consumption for Sherlovogorsk[a] ore fragmentation by the indicated methods

Serial No.	Fragmentation method	Size		Energy consumption, kW h/t
		Initial, mm	Final, mm	
1	Mechanical (commercial rod mill)	30	2	6.7
2	Mechanical (laboratory centrifugal mill)	30	2	11.7
3	Mechanical (laboratory roll mill)	30	2	12.6
4	Electrodischarge	30	2	13.1
5	Electrohydraulic	30	2	117.2

[a]The Sherlovogorsk deposit is located in the Lake Baikal region

pollution in the process of destruction. These advantages are provided by contactless and selective action on the material of "the working tool"—shock waves from the discharge channel. The accumulated experience on application of the EH method for RCP fragmentation demonstrates that it can also be widely used in construction industry. Experience accumulated in the destruction of rejected RCP demonstrates that the specific energy consumption for the destruction of PGK-59-12 hollow core slabs and reinforced concrete cross-beams and cross-pieces does not exceed 15–20 kW h/m^3. Concrete is completely separated from the armature and fragmented into rubble with sizes of 5–20 mm. The armature (except fine connecting one) is not deformed and is suitable for recycling in production. The calculated productivity of one installation is 12 thou and m^3 rubble per year, and the recoupment period is 1.5 years.

9.2 Electrodischarge Method of Fragmentation of a Concrete Filler

The electrodischarge technology that can rightly be considered as the next step in the development of electrophysical methods of solid material destruction is attractive not only for well drilling, fragmentation and disintegration of mineral raw materials, but also for fragmentation of the RCP filler due to its advantages in addition to those of the EH technology, including

- Improved selectivity of fragmentation and disintegration of materials with non-uniform structure; this advantage is especially pronounced for concretes that belong to multicomponent materials containing structural elements with different sizes and physical properties;
- The destruction rate and energy consumption are mainly determined by the electrophysical and elastic properties of materials and depend weakly on their strength and abrasion; the last property is important for fragmentation of high-brand concretes;
- The main technological parameters can be regulated in very wide limits. The parameters of the spark channel and perturbations it produces in a solid are sufficient for RCP recycling;
- Energy consumptions for fragmentation of materials by the ED method are comparable to mechanical fragmentation methods (Table 9.2) and are 10 times less compared to the EH method.

By the present time, the methods of calculating the parameters of the main components of ED installations with controllable end-product characteristics intended for fragmentation and disintegration of rocks and artificial solid materials have been developed. These methods in the adapted form can also be used for the development of technology of RCP filler fragmentation. However, to estimate the level of energy consumption for fragmentation of concrete pieces, an additional cycle of experimental researches was required because the material was treated not only in the initial state (having the RCP structure) that differed from rocks, but also in the state preliminary exposed to high-power electric discharges during preceding operation of RCP destruction. Therefore, reduced energy consumption for filler fragmentation in comparison with that for fragmentation of rocks can reasonably be expected.

Considering that the specific energy consumption for ED fragmentation is very sensitive to the parameters of energy package, methods of collecting the destructed product in the active zone, and extraction from it of ready fractions, it is necessary to elucidate experimentally the main directions of their minimization.

In this case, it is important to establish the relative contribution to fragmentation of two effects: electrohydraulic one produced by the discharge channel in water, and electrodischarge one produced by the discharge channel in concrete. It is determined by relationships between the probabilities of channel development in the thickness of concrete and in water and between the lengths of the channel parts in concrete and water. For this reason, a search for optimal parameters of energy sources and working parts of the installation for fragmentation minimizing the specific energy consumption was required.

To estimate the quality of fragmented products, a physicotechnical analysis of rubble from crushed concrete was performed that allowed the conclusion about its suitability to be drawn in the form of coarse filler for new concrete and reinforced concrete products.

9.2.1 Fragmentation Plant

In experiments, the installation was used whose technological scheme is shown in Fig. 9.1. To expand the interval of the examined energy deposition and to search for the optimal parameters of the working parts of the installation intended for fragmentation, IVGs with impulse amplitudes from 120 to 450 kV, impulse front duration of the order of 0.5×10^{-6} s, and deposited energy from 1.2 to 9.3 kJ were used; they were equipped with technological chambers of two types: with flat electrodes-classifiers or electrodes-classifiers in the form of paraboloide of rotation. At the top of the chamber, high-voltage electrodes were fixed. At its bottom, the replaceable grounded electrodes-classifiers were established. The ready product was collected under the electrode-classifier. The chambers were connected to a water supply system by means of sleeves and valves. The high-voltage electrode was shaped as a polyethylene cylinder 80 mm in diameter and 1500 mm in length and a metal core 20 mm in diameter connected to it. The grounded electrode-classifier was shaped as a flat disk with thickness of 20 mm and calibrated holes intended for removal of fragmentation products of concrete.

In flat electrodes, there were holes of different diameters (5, 10, 20, and 40 mm), and their number was chosen to provide identical porosities of sieves. The porosity of flat electrodes was approximately 0.4. To improve the placeability of the product passed through electrodes-classifiers, calibrated holes were drilled from the back side. The electrodes-classifiers of welded design with shapes close to paraboloide of rotation were also used in experiments, Fig. 9.2.

The forming cores (4 or 6 in number) were equally spaced along the electrode perimeter. Metal rings were welded on the cores. The distance between the rings was chosen to provide fragmentation of concrete to sizes less than 20, 40, or 70 mm. Such design provided good selection of fragmented product under the high-voltage electrode. The high porosity of these electrodes (at a level of 0.7–0.8) allowed the ready product to be extracted from the fragmentation zone before over-fragmentation.

The distance between the high-voltage electrode and the electrode-classifier was regulated using a fixing device clamped on the cap of the technological chamber.

Model samples and pieces of concrete subdivided into fractions obtained after extraction of reinforcing cage from sub-standard RCP, including wall slabs and balcony slabs (manufactured by the Tomsk plant of large-panel housing construction), were investigated. The cage was extracted on the installation described in Sect. 8.3. In this case, the energy consumed for destruction of each product and for auxil-

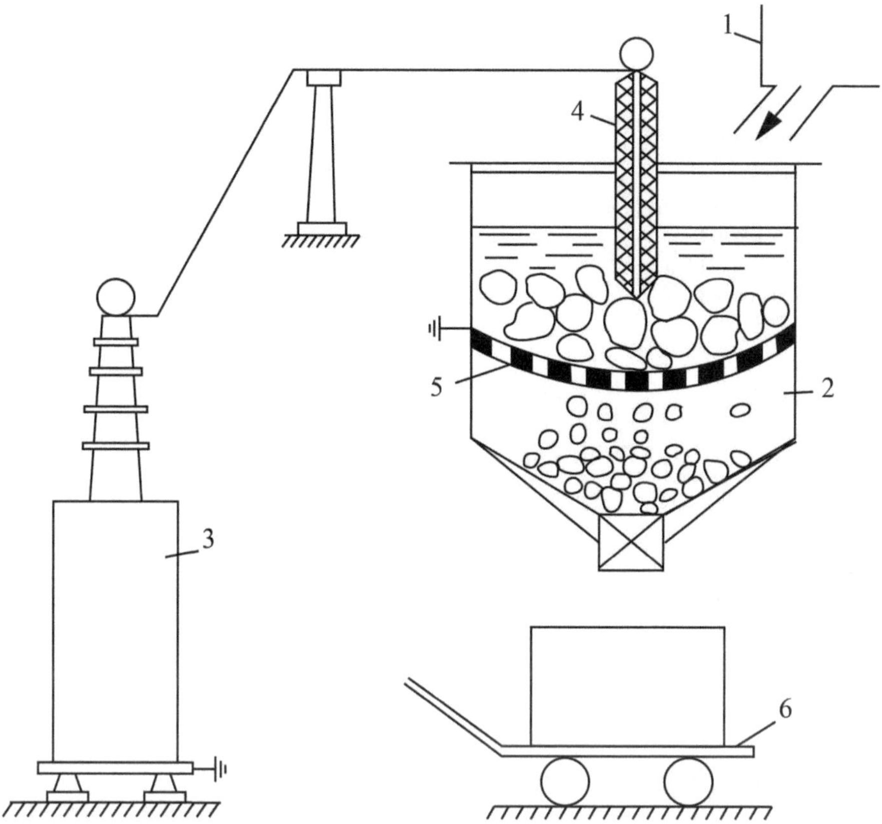

Fig. 9.1 Technological scheme of ED fragmentation

Fig. 9.2 Photographs of electrodes-classifiers of welded design

iary operations and also the size distribution of concrete fragments were measured. Results of experiments are presented in Table 9.3.

Table 9.3 Fragmentation parameters of concrete extracted from RCP of two types

Parameters	Internal wall slab	Balcony slab
Product volume, m^3	0.31	0.32
Product mass, kg	668	815
Metal mass, kg	30.5	33.3
Specific energy consumed for destruction, kW h/t	4.2	9.0
Fractional composition of fragmented concrete, %		
0–5 mm	6.3	9.6
5–10 mm	8.4	13.7
10–20 mm	17.3	24.6
20–40 mm	13.5	20.0
40–70 mm	12.0	13.9
Exceeding 70 mm	42.5	18.2

From the Table it follows that to obtain the fraction 0–40 mm, more than 54% of concrete produced by destruction of the internal wall slab and 32% of concrete produced by destruction of the balcony slab should be repeatedly fragmented.

9.2.2 *Specific Energy Consumed for Fragmentation*

In the process of concrete fragmentation by the ED technology the factors influencing on the energy parameters of the process are about the same as those of rock fragmentation, including

(1) Energy source parameters:

- Energy per impulse that can be varied by changing the discharge capacitance of the storage device and its voltage and the inductance of the discharge circuit,
- Rate of increase of the voltage applied to the object (steepness of the impulse front),
- Time of energy liberation in the discharge channel,

(2) Length of the working gap,
(3) Initial size of the destructible product.

The parameters of energy sources, along with the properties of destructible body, determine the law of energy liberation in the discharge channel and hence the energy consumed for the destruction process and its productivity. The relationship between the source parameters and the law of energy liberation in breakdown of solids was investigated in Sect. 4.2 where it was demonstrated that the resistance of the breakdown channel is approximated fairly well by expression (4.9). The relationships between the spark constant and the properties of the destructible body are described by dependence (4.13).

Values of the spark constant for concrete of different brands are presented in Table 4.5.

For the *R*–*L*–*C* discharge circuit with R_{ch} given by Eq. (4.9), it is expedient to use the parameter given by Eq. (4.18).

Then the energy liberated in the discharge channel during the first half-period of the discharge current $T_1 = \pi.\omega^{-1}$ (as is well known, it makes the decisive contribution to the destruction) will be determined by formula (4.31).

By way of example, we now calculate the energy liberated in the discharge channel to destruct brand 200 concrete. The IVG parameters were $U_0 = 357$ kV, $C = 0.094$ μF, $L = 150$ μH, the channel length $l_{ch} = 0.1$ m (set equal to the interelectrode gap *S*), and the spark constant for brand 200 concrete $A_{ch} = 290$ V $s^{0.5}$ m^{-1}. The energy stored in the storage device is

$$W_0 = \frac{CU_0^2}{2} = \frac{0.094 \cdot 357^2}{2} = 6000\,\text{J}$$

Then

$$CS = 0.042,$$

$$I_{\text{max}} = 1.5(0.67 - CS) \cdot U_0\sqrt{\frac{C}{L}} = 8.42\,\text{kA},$$

$$D = 1.35\,\pi\,CS = 0.18\,\text{rad},$$

$$T_1 = \pi\sqrt{LC} \cdot (1 + 1.35CS) = \pi \times 3.75 \times 10^6\ \text{s},$$

$$\omega = 1/\sqrt{LC} = 1/3.75 \times 10^{-6} = 0.267 \times 10^6\ \text{s}^{-1},$$

$$W_{ch} = 290 \times 0.1 \times 8.42 \times 10^3 \times 0.267^{-0.5} \times 10^{-3} \times [2\pi - \sin(2\pi + 0.36)]^{0.5} = 1152\ \text{J},$$

which yields

$$\eta = \frac{W_1}{W_0} = \frac{1152}{6000} = 0.192.$$

Hence, for an energy of 6 kJ stored in the source, only the portion of energy equal to $W_{ch}/W_0 = 0.192$ will be liberated in the discharge channel.

To increase this portion, the criterion Π sensitive to the ratio l_{ch}/U_0 must be increased. However, this ratio is limited by the electric strength of destructible concrete. In the breakdown of samples of brand 200 concrete with thickness of 0.20 m on the front of the impulse of positive polarity with steepness of 1050 kV/μs, the electric strength (the average breakdown stress U_{br}/S) was 14 kV/cm. This means that the ratio S/U_0 could not exceed 0.714×10^{-6} m/V. In this limiting case, $U_0 =$ 280 kV, $CS = 0.107$, $W_0 = 684$ J, $I = 5910$ A, $\omega = 0.267\ \mu s^{-1}$, $W_{ch} = 1550$ J, and the ratio $W_{ch}/W_0 = 0.42$.

Results of calculations were tested in a series of experiments. Pieces of concrete with sizes from 20 to 40 mm and total mass of 57 kg were loaded into the discharge chamber. The electrode-classifier represented the plane with holes 20 mm in diameter. The interelectrode gap was 100 mm. The amplitude of impulses having frequency of 1 impulse/s was changed when going from one series of experiments to another. Water from the discharge chamber was removed, and the output product was collected in the storage vessel arranged under the sieve and then weighed. The specific energy consumption was calculated from the output product. Results of experiments are presented in Table 9.4.

It can be seen that the specific energy consumption is sufficiently critical to the energy of the impulse source and has a minimum at $W_0 = 4.8$ kJ.

To investigate the influence of the interelectrode gap on the specific energy consumption, experiments were performed for interelectrode gaps of 85, 100, 150, and 200 mm using the electrode-classifier with the distance between the rings equal to 40 mm, Fig. 9.2. In the experiments, 150 kg samples were used. The size of the destructible product changed from 40 to 70 mm. The energy accumulated by the IVG was 3.5 kJ. Experimental results are given in Table 9.5.

It can be seen that the dependence $W_{spes} = f(S)$ has a minimum, that is, there are optimal relationships between the parameters of the discharge circuit and the interelectrode gap length. Such character of the dependence $W_{spes} = f(S)$ can be explained by low energy consumption during the first impulse half-period for small interelec-

Table 9.4 Influence of the IVG parameters on the specific energy consumed for concrete fragmentation

IVG parameters				Number of impulses	Mass of the product collected under the sieve, kg	Specific energy consumption, kW h/t
U_0, kV	L, μG	C, μF	W_0, kJ			
272	150	0.094	3.4	300	9.9	30
318	150	0.094	4.8	300	32.6	15
357	150	0.094	6.5	300	21.5	25

Table 9.5 Influence of the interelectrode gap on the specific energy consumed for concrete fragmentation at constant stored energy of 3.5 kJ and initial product size 70–150 mm

Serial No.	Interelectrode gap, mm	Specific energy consumption, kW h/t
1	85	4.1
2	100	3.2
3	150	3.3
4	200	5.2

trode gaps and reduced probability of breakdown channel penetration into a solid for large S when the impulse amplitude becomes insufficient for breakdown of the solid. The electrodischarge fragmentation regime is transformed into the electrohydraulic one. The energy levels used in the electrodischarge technology are insufficient for effective electrohydrauic destruction, and disintegration of the solid stops. The dependences $W_{spes} = f(S)$ have identical characters for all materials.

The influence of the size of the initial product on the specific energy consumed for concrete fragmentation was investigated using the sieve-classifier with 40 mm holes. The initial material comprised two fractions—40–70 mm and 70–130 mm. The material was destructed in the chamber with interelectrode gap of 150 mm. The energy stored in the IVG was 3.5 kJ, and the number of applied impulses was 300. The specific energy consumption was determined from the weight of the fragmented product and the consumed energy equal to 1.62 kW h/t for fraction 40–70 mm and 3.3 kW h/t for fraction 70–150 mm. The consumed energy was directly proportional to the size of the destructible product.

The investigations performed were insufficient for exhaustive elucidation of the dependence of the fragmentation process parameters on the joint influence of different factors. Further search is required of optimal relationships between the complex of factors and the specific energy consumption in the process of ED fragmentation of concrete.

9.2.3 Granulometric Characteristics of Products

The particle size distribution of the final product obtained in different regimes of energy deposition and for different sizes of initial raw materials were investigated. The initial product having a mass of 120 kg was loaded with water into the fragmentation chamber 600 mm in diameter, Fig. 9.3.

After application of a series of impulses, water was removed from the chamber. Crushed concrete was subdivided into fractions using a standard system of sieves, Fig. 9.4.

To estimate the suitability of rubble from crushed concrete, its physicotechnical properties were investigated.

Table 9.6 Results of investigations of particle size distribution for crushed concrete

No.	Characteristic	Serial number of experiment					
		1	2	3	4	5	6
1	Size of holes in the sieve-classifier, mm	20	40	40	40	40	40
2	Interelectrode gap, mm	80	100	100	150	200	100
3	Sizes of fractions of the initial product, mm	40–70	40–70	70–150	70–150	70–150	40–70
5	Weight of the product collected under the sieve-classifier, kg	16.4	69.7	30.0	29.0	18.7	36.7
6	Energy per pulse (disregarding losses), J	3500	3500	3500	3500	3500	6500
7	Voltage amplitude, kV	272	272	272	272	272	357
8	Number of pulses	100	100	100	100	100	100
9	Relative content of fractions in the final product, %						
	from 0 to 5 mm	13.4	5.2	7.0	7.6	7.9	10.4
	from 5 to 10 mm	21.4	6.6	11.6	14.1	14.6	13.9
	from 10 to 20 mm	65.2	14.0	30.0	32.4	34.5	32.1
	from 20 to 40 mm	–	74.2	51.4	45.8	43.0	43.6

Fig. 9.3 Initial fragmentation product

Fig. 9.4 Appearance of the final product of secondary crushing of concrete results of investigations are presented in Table 9.6

9.2.4 Physical and Technical Characteristics of a Crushed Concrete Filler

In the construction industry for the production of heavy concrete gravel or crushed stone, which are mixtures of grains of natural stone with a size of 5 to 70 mm, are used.

A mixture of pellet grains formed as a result of aeration of rocks is called *gravel*. A mixture consisting of angular grains with rough surface obtained by fragmentation of ejected volcanic, sedimentary, or metamorphic rocks is called *rubble*.

In the process of fragmentation of sub-standard concrete products, the crushed product can be characterized as rubble though it comprises gravel extracted in its original form.

To study the properties of rubble from crushed concrete used to estimate its quality, it is necessary to determine:

(1) Grain composition,
(2) Contents of needle-shaped and lamellar grains,
(3) Bulk density,
(4) True density,
(5) Contents of dust and clay particles,
(6) Content of organic impurities,
(7) Content of crushed grains,

(8) Crushability under compression in the cylinder,
(9) Abrasion.

Speaking about the quality of rubble, the main properties which the filler should possess must be established. First of all, the coarse filler in concrete should have rational grain composition related with intergrain voidness that determines the placeability of concrete. This allows high-quality concrete to be produced with minimal cement consumption. At the same time, the presence of needle-shaped and lamellar grains in the filler leads to increased intergrain voidness, worsens the placeability of the concrete mixture, and reduces the strength of concrete. The purity of the filler affects significantly the strength of concrete. The dust clay impurities form a film on the filler grain surface which prevents grain coupling with cement stone and requires an increased consumption of water. The compressive strength characteristic of rubble brand is the filler crushability; an important property of the filler is also its abrasion. The lower the abrasion of the coarse filler, the higher is the resistance of road concrete with such filler to the action of road transport.

Physicomechanical properties of the filler from crushed concrete determined from tests standardized by the state (GOST 8269-76, GOST 10268-80 and GOST 8268-82), are presented in Table 9.7.

The content of dust and clay particles was less than 1% for all fractions. It seems likely that this was due to the fact that concrete was crushed in water. As a result, crushed rubble was washed out from clay particles.

Tests on detection of organic impurity demonstrated that the filler was suitable for production of concrete. Suitability of rubble for various applications of concrete was estimated in accordance with U.S.S.R. Standard (GOST 10268-80).

In the process of production of heavy concrete, rubble from crushed concrete is suitable only for production of brand 250 concrete and of concrete with lower brands.

Table 9.7 Results of testing of rubble from crushed concrete

No.	Rubble characteristics	Sizes of fraction, mm		
		5–10	10–20	20–40
1	Content of needle-shaped and lamellar grains, %	24.0	6.0	19.4
2	Bulk density, kg/m^3	1170	1300	1155
3	True density, kg/m^3	2440	2440	2440
4	Intergrain voidness, %	52.0	46.7	53.0
5	Crushability, %	16.3	12.7	21.7
6	Abrasion, %	30.2	26.0	35.0

High crushability of fractions 5–10 and 20–40 mm is a disadvantage. The fraction 10–20 mm having crushability FR12 can be used for preparation of M350 and M300 concrete. This conclusion was made from testing of samples of concrete with coarse filler of fraction 10–20 mm and of reference samples of the same sizes using gravel rubble (cubic samples with 10 cm edge).

In the experiments, portland-cement was used.

Initial materials comprised (per 1 m^3):

- Cement 420 kg,
- Small filler—quartz sand—590 kg,
- Coarse filler 1237 kg,
- Water: 188 L for samples with natural filler (with water/cement ratio W/C = 0.45) and 218 L for samples with filler from crushed concrete (W/C = 0.52). Samples were steamed in a laboratory steaming chamber. Compressive tests according to U.S.S.R. Standard (GOST 10180-78) were performed in a day. Results of testing are presented in Table 9.8.

The mobility of the concrete mixtures prepared from crushed concrete and natural filler can be leveled by changing of the W/C ratio (the slump cone of the mixture with the secondary filler at W/C = 0.45 was close to zero). From Table 9.8 it follows that the strength of concrete from crushed rubble is not worse than that of concrete from natural filler.

Technical requirements to rubble for construction works in comparison with the test results allow the conclusion to be made on the applicability of rubble from crushed concrete for these works, because rubble from crushed concrete can be referred to the improved group against the criteria of grain shape, and the fraction 10–20 mm can be referred to the cubic one. The crushability for different rubble fractions was Dr16 for fraction 5–10 mm, Dr12 for 10–20 mm, and Dr24 for 20–40 mm, respectively. The abrasion for fractions 5–10 and 10–20 mm corresponded to grade I–P, and for fraction 20–40 mm it corresponded to grade I–Sh.

The grain structure of rubble intended for different kinds of works was adjusted by individual dosing of different fractions. The content of different fractions in coarse filler depended on the composition of concrete. For ED technology of secondary crushing of concrete, the required grain composition can be regulated by adjusting the energy deposited into the discharge channel.

Table 9.8 Results of comparative tests of samples of concrete with filler from crushed concrete and with natural filler

Brand of concrete	Type of filler	Compressive strength, MPa	Density, kg/m^3
M300	Crushed	26.1	2240
M300	Natural	25.5	2340

Table 9.9 Filler yield depending on the model type and the destruction regime

Model type	Filler fractions, mm	Strength of the sand-cement mixture, 10^7 Pa	Destruction regime	W_0, J	T, μs	F, %
1.	8.8–12.5	1.9	1	4680	6.4	7.0
2	17.7–25.0					12.8
3	25.0–35.0					22.1
4	8.8–12.5 and 17.7–25.0					9.2 and 16
5	8.8–12.5 and 25.0–35.0					5.4 and 12.1
6	12.5–17.7	1.9	2	3170	6.4	6.9
			3	4680		12.0
			4	6240		13.2
7	12.5–17.7	2.7	5	3330	4.8	3.6
			6		15.1	4.8
			7		21.4	3.7

9.3 Extraction of Coarse Filler from Concrete

In the process of recycling sub-standard RCP, the problem arises of the most complete extraction of the filler from the formed pieces of concrete for their repeated application in new products. The filler yield and the resulting size distribution of destruction products were studied depending on the composition of concrete and the parameters of energy deposition into the discharge channel. To study the influence of the composition of concrete on the size distribution of the formed fragments, experiments were performed with model samples of concrete having sizes 10 × 10 × 10 cm. A total of 7 samples (Table 9.9) were prepared using gravel filler with compressive strength of 1.1×10^8 Pa and content of 51%. For samples containing the filler with two fractions, their specific weight contents were chosen identical. The sand-cement binding mixture had the sand-cement ratio 1:1 and the water/cement ratio of 0.3.

The breakdown of samples of concrete was initiated by single voltage impulses in water; the "tip-plane" electrode system was used. In this case, the energy W_Σ liberated in the breakdown channel during time t of discharge current flow and the energy W_1 liberated during the first impulse half-period were calculated from the measured discharge current and the voltage drop across the spark channel. For each combination of experimental conditions, breakdown was initiated in 8–10 samples. After destruction of concrete and drying of sample fragments, a sieve analysis was performed using the standard sieve system.

Figures 9.5 and 9.6 show the average sizes of the destruction products for the impulse energy $W_0 = 4680$ J and the ' period $T = 6.4$ μs (regime I in Table 9.9).

Two maxima can be distinguished on the dependences shown in Fig. 9.5. The first maximum is determined by the initial sizes of the filler—the strongest component of the concrete samples, and the second maximum is connected with the geometrical sizes of the sample destroyed only partly at the chosen level of energy. Thus, the yield F of the filler with sizes 8.8–12.5, 17.7–25, and 25–35 mm was 7, 12.8, and 22.1%, respectively, which corresponded to 13.7, 25.1, and 43.35% of filler content in the sample before destruction. Since the degree of crushing of the samples characterized, for example, by the average size of fragments in the chosen energy regime exceeded the size of the filler fraction, it was natural that its maximum yield corresponded to a larger fraction. For samples with two initial fractions of the filler, the maxima of the size distribution corresponding to each fraction were observed, Fig. 9.6.

The influence of the impulse energy for $T_1 = 6.4$ μs on the results of destruction was investigated for samples containing filler fraction 12.5–17.7 mm. The maxima in the average size distribution were observed for the initial fraction irrespective of the amount of energy, Fig. 9.7.

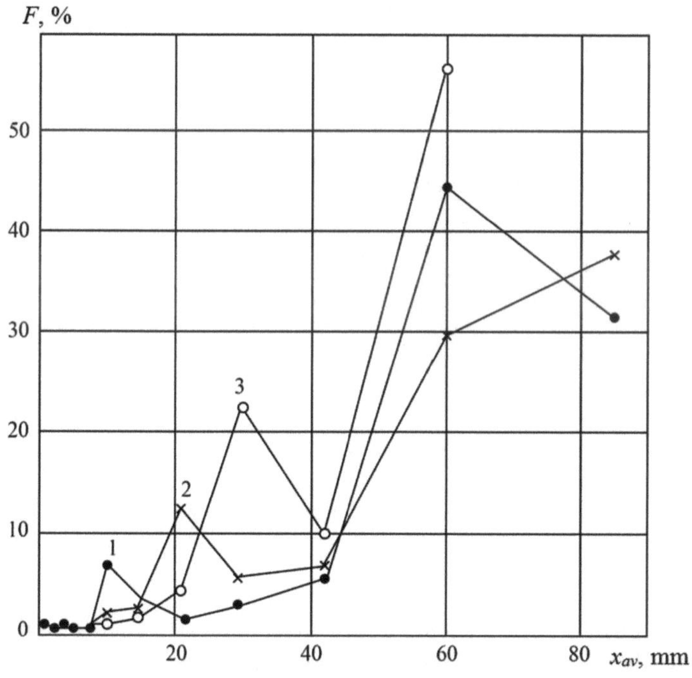

Fig. 9.5 Average sizes of fragments of concrete comprising single initial fraction of the filler. Here curve 1 is for fraction 8–12.5 mm, curve 2 is for fraction 17.7–25 mm, and curve 3 is for fraction 25–35 mm

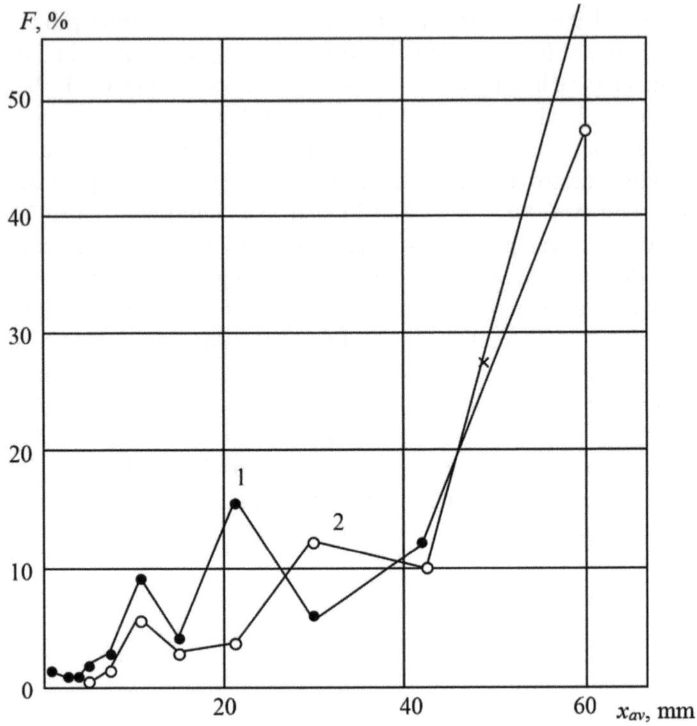

Fig. 9.6 Average size distribution of the crushed concrete fragments with the filler comprising two initial fractions. Here curve 1 is for fractions 8–12.5 and 17.7–25 mm and curve 2 is for fractions 8–12.5 and 25–35 mm

Table 9.10 Size distribution of rubble from crushed concrete

Diameters of holes in the blocking sieves, mm	40.0	20.0	10.0	5.0	2.5	1.25	0.63
Mass remaining on the sieves, %	6.4	16.6	55.5	14.2	1.7	1.4	0.8

A comparison of the filler yields in destruction regimes 2, 3, and 4 from Table 9.10 demonstrates that the increase of the impulse energy causes the filler yield to increase.

However, the specific yield per unit energy decreased when the deposited energy achieved a certain critical value. For the chosen regime, it was $W_0 \sim 5$ kJ.

Taking into account the wave character of loading during pulsed electric breakdown of solids and different acoustic properties of the filler and binder, the influence

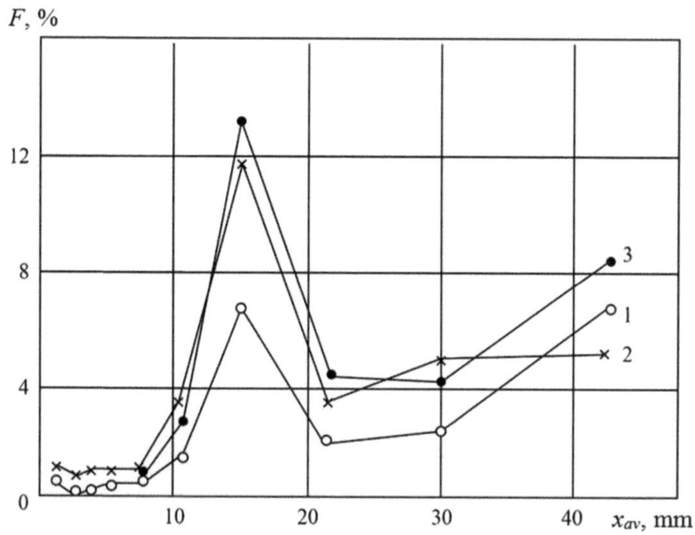

Fig. 9.7 Average size distribution of concrete particles in the range from 0 to 45 mm. The filler fraction was 12.5–17.7 mm. Here the destruction regimes of the curves correspond to models presented in Table 9.10: curve 1 is for model 2, curve 2 is for model 3, and curve 3 is for model 4

of energy deposition period and, as a consequence, the time characteristics of the mechanical stress field on the filler yield should be expected. By the present time, the influence of the loading period on the uniformity of rock and ore crushing, selectivity of destruction, and degree of crushing of inclusions has been investigated both for explosives and for the electric discharge. Essential influence of the time dependences of loads on the above-indicated characteristics was noted.

To study the influence of dynamics of energy liberation in the breakdown channel in concrete on destruction, two series of experiments were performed with impulse energy of 3330 and 4440 J, respectively. In the first series of experiments, $U_0 =$ 400 kV and the discharge capacitance $C_{disch} = 0.0417$ μF; in the second series of experiments, $U_0 = 390$ kV and $C_{disch} = 0.0583$ μF. The inductance L changed in both series of experiences from 10 to 400 μH. The period T_1 of the discharge current application increased from 4 to 26 μs. Samples of concrete with filler sizes 12.5–17.7 mm (model 7) were used in the experiment.

The time $T_1/2$ was normalized to the time during which the longitudinal sound oscillations passed through the filler particle ($t = d/s_0$, where d is the average linear size of the filler particle and c_0 is the sound speed in it), Fig. 9.8. In the experiment with $d = 15.1$ mm, $c_0 = 3950$ m/s, and $T_1 = 15.1$ μs, the maximal filler yield was 4.8% in the first series and 5.6% in the second series of experiments, respectively. The sizes in the range from 0 to 45 mm in the first series of experiences for energy liberation regimes Nos. 5, 6, and 7 from Table 9.10 are shown in Fig. 9.9.

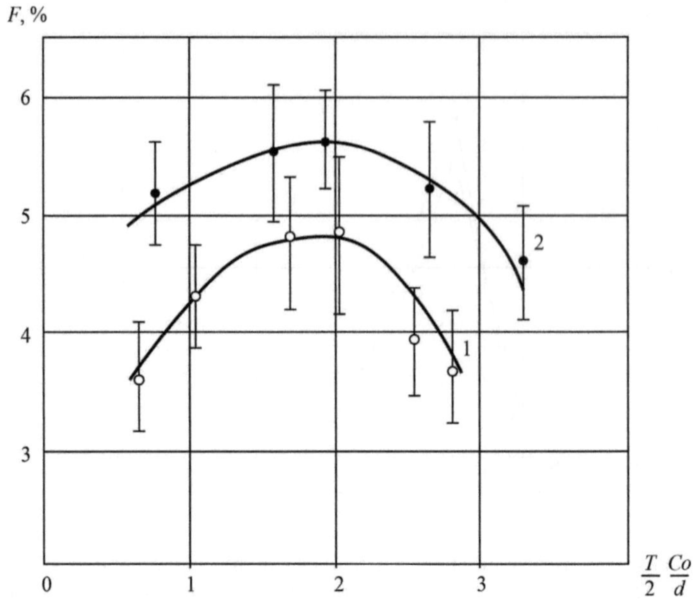

Fig. 9.8 Dependence of the filler yield on the energy deposition time ($T_1/2$) for impulse energy of 3330 (curve 1) and 4440 J (curve 2)

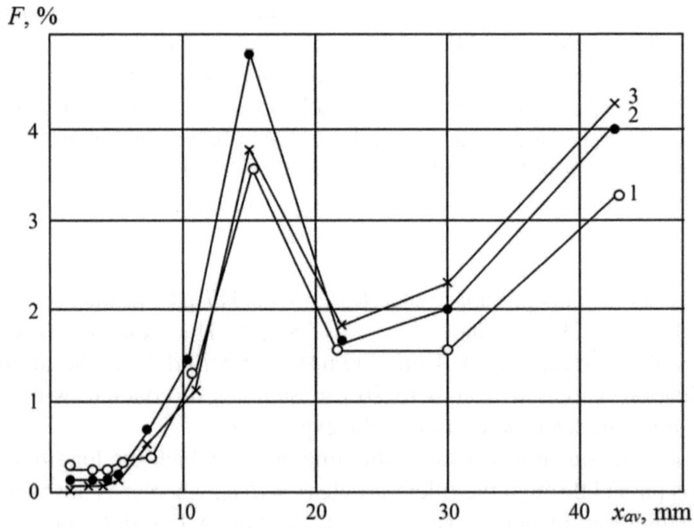

Fig. 9.9 Average size distribution of particles of concrete for different energy liberation regimes. The filler fraction 12.5–17.7 mm and the destruction regimes Nos. 5 (curve 1), 6 (curve 2), and 7 (curve 3) according to Table 9.10

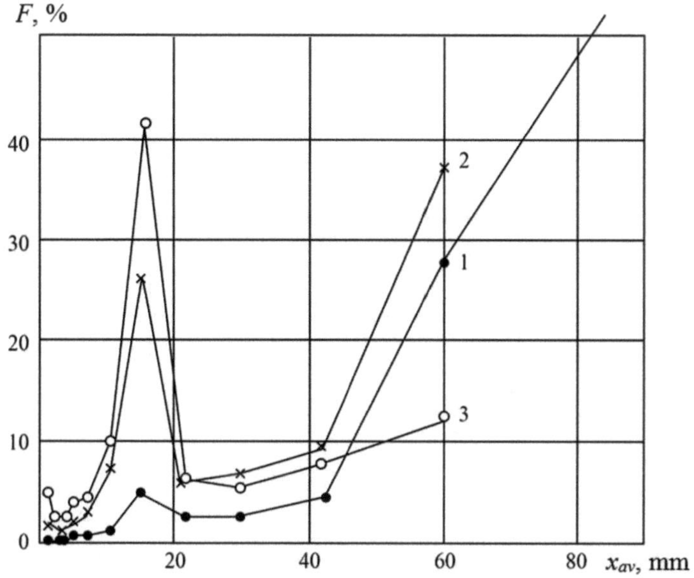

Fig. 9.10 Influence of the number of applied voltage impulses on the particle size distribution for concrete destructed in regime 6 for models 1 (curve 1), 3 (curve 2), and 5 (curve 3) from Table 9.10

Results of experiments demonstrated that there existed an optimal time of energy liberation at which the filler yield was maximal; it was approximately equal to the time of double wave passage through the filler particle.

The level of the filler yield decreased with increasing strength of the binding mixture, which follows from a comparison of destruction regimes with close energy values (regimes 2 and 5 in Table 9.10).

The filler yield can be increased significantly by application of several high-voltage impulses to the sample of concrete. Thus, under application to the sample of a single impulse with the parameters corresponding to regime 6 from Table 9.10, the filler yield was 4.8%; under application of three impulses, it increased to 26.6%; and under application of five impulses, it reached 41.3%. In the latter case, ~80% of the filler contained in the sample was extracted. The particle size characteristics for the examined regimes are illustrated by Fig. 9.10.

During fragmentation of solids, the size distribution of fragments was typically described by special mathematical functions, for example, of the Rosin–Rammler type.

For analytical representation of the curves illustrating the particle size distribution for concrete, the dependences shown in Figs. 9.5 and 9.6 were generalized; they are shown in coordinates $\ln\ln F^{-1}$ and $\ln x$ in Fig. 9.11.

The typical curve with filler of one fraction has three characteristic segments approximated as follows:

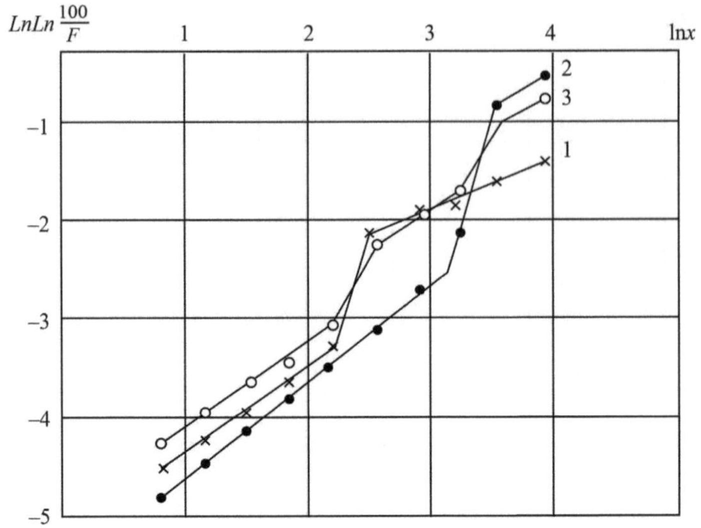

Fig. 9.11 Particle size distributions for filler fractions 8.8–12.5 mm (curve 1), 8.8–12.5 mm (curve 2), and 25–35 mm (curve 3)

$$
\begin{aligned}
& n1\ln x + \mathrm{C1} \text{ for } \ln x \leq \ln b1, \\
& \ln\ln F^{-1} = n2\ln x + \mathrm{C2} \text{ for } \ln b1 \leq \ln x \leq \ln b2, \text{ and} \\
& n3\ln + \mathrm{C3} \text{ for } \ln x \geq \ln b2.
\end{aligned}
$$

Here C1, C2, and C3 are constants and *b1, b2, b3, n1, n2, and n3* are the distribution parameters.

The high yield of fragments having sizes of the filler fraction corresponded to the second segment.

The size distributions of fragments for samples with filler containing several fractions can be represented analogously. In this case, the yield of each fraction will correspond to its own segment on the curve of particle size distribution.

The results obtained allow the main conclusion to be drawn that the ED technology can compete with mechanical one for intermediate and small fragmentation of concrete due to a number of factors, including

- Specific energy efficiency comparable with the mechanical methods and exceeding that of the EH method by an order of magnitude;
- The destruction in technical water simultaneously provides washing from clay particles and cement stone;
- Activation of the surface of products in the course of ED fragmentation provides enhanced strength of secondary concrete with rubble from crushed concrete as a filler;
- In the process of destruction of concrete, the probability of the yield of fractions corresponding to sizes of the initial filler particles was higher than that of newly

formed fractions; this allowed the initial filler to be efficiently extracted from RCP destruction products;

- The filler yield depended on the impulse energy and energy deposition time; moreover, its maximum yield under application of a single pulse corresponded to energy of approximately 5 kJ deposited into the discharge gap and the impulse duration approximately equal to the time of double passage of mechanical stress waves through the filler particles. Realization of such regime is not difficult;
- The energy consumed for fragmentation of concrete using flat sieve-classifiers was 15–30 kW h/t, and it amounted to 3–5 W h/t when using parabolic sieve-classifiers;
- Rubble from crushed concrete could be used to produce heavy brand 250 concrete and concrete of lower brands, and the extracted primary rubble of fraction 10–20 mm can be used for production of brand 350 concrete.

9.4 Properties of Concrete with Crushed Stone Filler from Recyclable Reinforced Concrete

To study the properties of concrete with rubble filler extracted from utilized RCP, rubble of crushed concrete with fraction 0–40 mm was used. The rubble size composition is presented in Table 9.10.

Rubble from crushed concrete represents fragments of grains of natural filler covered with a cement stone film. As the rubble grain size decreases, the amount of cement stone increases. Sandy grains represent a mixture of sand and cement stone grains and of the solution. Grains of the filler larger than 20 mm include a considerable part of fragments of concrete. The presence of the cement stone in the filler and concrete fragments requires increased amount of water for preparation of the mixture of concrete. Therefore, rubble is classified by the fractions: exceeding 20 mm, 5–20 mm, and smaller than 5 mm. To prepare concrete, rubble with fraction 5–20 mm was used as large filler; its particle size distribution and main physicomechanical characteristics are presented in Table 9.7.

As a binder, portland cement with chemical composition presented in Table 9.11 was used.

To prepare the concrete mixture, natural quartz-feldspar sand as small filler and tap water were used. The composition of concrete was selected against the criterion of the least voidness of the filler mixture. The properties of concrete mixtures and of concrete based on rubble from crushed concrete with natural sand used as small filler and screening of concrete fragments less than 5 mm were estimated. The compositions and properties of concrete mixtures are presented in Table 9.12.

Table 9.11 Chemical composition of cement

SiO_2	Al_2O_3	F_2O_3	CaO	MgO	SO_3	Σ
22.1	6.4	3.6	64.5	1.1	0.2	97.9

It was established that concrete mixtures based on screening of crashed stone require larger (by 10–20 L/m^3) amount of water to produce mixtures of concrete convenient for placing compared with mixtures based on natural send. This is explained by the composition of screened crushed stone and its higher (approximately by 11.5%) water absorption. In addition, visual inspection of grains testified to their highly developed surface. The strength characteristics of concrete with screened crushed stone were by 15.6–21.2% lower than those of concrete based on natural sand. Thus, the application of screened crushed stone as small filler is inefficient due to a decrease in the strength and increase in the required amounts of water and cement.

The increased consumption of water and binder and the presence of rubble of cement stone on the surface of grains caused the creep of concrete to increase and the elasticity modulus and prismatic strength of concrete to decrease [8]. One of the important methods of improving the quality of concrete with such filler is modification of impurity rubble. The possibility of modification is provided by the presence of new hydrate cement formations on the cement surface. Among the well-known methods of modification (mechanical, chemical, and thermochemical ones), the most technological is the chemical method. During testing, supersoftener C-3 and acetic acid were used as modifiers. The choice of modifiers allowed hydrophobic-hydrophilic properties of the rubble surface to be changed in the required direction. The choice of acetic acid was caused by its ability to interact with portland filler grains forming

Table 9.12 Compositions and properties of concrete mixtures

Content of components, kg/m^3					Cone slump, cm	Density of concrete mixture, kg/m^3	Limiting compressive strength, MPa, with age, days	
Rubble	Sand	Screened crushed concrete	Cement	Water			7	28
1040	675		385	200	3	2300	18.0	25.7
1010	690		405	210	5	2310	17.8	24.7
980	710		425	220	9	2330	17.6	22.7
1060	670		310	200	4	2240	16.5	22.3
1040	680		325	210	6	2260	15.8	21.6
1000	720		340	220	11	2240	17.7	21.2
1080		480	405	210	2	2130	16.3	21.3
1060		505	425	220	4	2210	17.4	20.5
1040		520	445	230	7	2230	16.3	18.7
1090		500	325	210	3	2130	14.3	18.3
1060		520	340	220	5	2140	12.8	17.8
1040		540	355	230	9	2180	12.7	167

hydrocarbon chains of molecules. The latter, as a rule, possess a weak force field and are sliding planes.

During tests concrete with modified filler was placed on a metal tray. The preliminary prepared solution of additives-modifiers was sprayed over the filler surface. Then other components of concrete were added. The mixture was stirred for 3 min, and after that the properties of the concrete mixture were determined for 7- and 28-days aged samples—cubes 10 × 10 × 10 cm were prepared from concrete. Possible concentrations of C-3 solution (with concentration of 12 g/L) and acetic acid (with concentration of 5%) were preliminary determined. To process rubble, 1 kg of C-3 solution or 35 cm^3 of acetic acid solution was required. The physicomechanical properties of the concrete mixture and of concrete are presented in Table 9.13.

The results presented above confirmed that modification of rubble from crushed concrete using the supersoftener C-3 solution allows the placeability of the concrete mixture to be increased from 5 to 11 cm, the density of the concrete mixture to be increased by 3.7%, and the strength of concrete to be increased by 3.7%. The modification of rubble by the acetic acid solution allowed the density of concrete mixture to be increased by 3.2%, the placeability of the concrete mixture to be increased from 5 to 8 cm, and the strength of concrete to be increased by 32.1%.

It should be noted that modification of the rubble surface by solutions of acetic acid and C-3 contributed to the stabilization in time of the properties of the concrete mixture.

The data on the change of mobility of concrete mixture for the indicated period of time are presented in Table 9.14.

Thus, processing of the filler with modifying solutions allowed the mobility of the concrete mixture and its molding properties to be preserved until molding.

To determine the main operational characteristics of concrete prepared from different concrete mixtures, cube samples 10 × 10 × 10 cm of two types were prepared to determine the frost resistance, and beams 10 × 10 × 40 cm were prepared to determine the prismatic strength and the elasticity modulus. To compare the data, concrete mixtures with the same W/C ratios and the same mobilities of the concrete mixture were prepared from gravel class 5–20 mm. The composition of the concrete mixture is presented in Table 9.15.

Table 9.13 Properties of the concrete mixture and of concrete

Composition of the concrete mixture and consumption of components, kg/m^3				Water-modifier	Mobility, cm	Density of concrete mixture, kg/m^3	Limiting compressive strength, MPa, with age, days	
Rubble	Sand	Cement	Water				7	28
1010	690	405	210	–	5	2310	17.6	24.3
1010	690	405	210	C-3	11	2360	17.1	26.2
1010	690	405	210	Acetic acid	8	2410	23.1	32.1

Table 9.14 Change of the mobility of concrete mixture with time

Filler type	Modifier	Mobility of the concrete mixture, cm, for the indicated time period	
		Immediately after preparation	In 20 min
Rubble from crushed concrete	–	5.0	1.5
	C-3	11.0	9.0
	Acetic acid	8.0	7.0

Table 9.15 Composition of the concrete mixture

Consumption of components per 1 m^3, kg				Density of the concrete mixture, kg/m^3	Mobility of the concrete mixture, cm
Gravel	Sand	Cement	Water		
1280	570	340	175	2370	6.0

Table 9.16 Composition and properties of concrete

Filler type	Modifier	Cement consumption, kg/m^3	Limiting compressive strength for samples, MPa		Elasticity modulus, 10^{-3}	Mass losses after 15 cycles,%	Frost resistance, brand
			Cube	Prism			
Gravel	–	340	19.8	13.7	22.4	2.9	−200
Rubble from crushed concrete	–	405	24.3	16.5	24.8	3.7	−150
Rubble from crushed concrete	C-3	405	25.4	17.4	25.2	2.3	−200
Rubble from crushed concrete	Acetic acid	405	32.1	23.3	29.6	2/1	−200

After preparation, the samples were hold for 28 days under normal conditions (at a temperature of 18 ± 2 °C and humidity exceeding 95%). The prismatic strength of concrete was determined according to GOST 24452-80, the frost resistance was determined according to U.S.S.R. Standard (GOST 10060-87), and the elasticity modulus was determined according to U.S.S.R. Standard (GOST 24452-85) under loading of 0.3. The properties of concrete are presented in Table 9.16.

For rational choice of the composition, the functional characteristics of concrete based on rubble from secondary concrete were not inferior to those of concrete based on natural stone fillers; hence, it can be used for design of any type. The prismatic strength and the elasticity modulus of the modified filler increased. Thus when rubble was modified by the solution of acetic acid, the elasticity modulus of concrete reached 2.69×10^4 MPa, which corresponded to conventional concrete based on natural stone fillers.

The material presented in Chap. 9 allows us to draw the following conclusions:

- Rubble from crushed concrete with fraction 5–20 mm can be used to prepare concrete of brands up to M-300 inclusive, without restrictions on the application field;
- Small fractions of the crushed product (less than 5 mm) can be used to prepare a concrete mixture and concrete of brands up to M-200 with overuse of cement and subsequently employed for manufacture of unimportant designs;
- To increase the efficiency of application, it is recommended to modify the rubble from crushed concrete by C-3 or acetic acid solutions; the modification of the rubble surface allows the placeability of the concrete mixture to be increased, its "survivability" to be improved, and the strength of concrete to be increased;
- Concrete based on the modified filler possesses satisfactory strain characteristics and can be recommended for designs of all types;
- Concrete based on rubble from crashed concrete possesses sufficient frost resistance and is not inferior to concrete based on natural stone fillers.

References

1. Emelin MA, Morozov VN, Novikov NP et al (1990) New methods of rock destruction. Publishing House Nedra, Moscow, 240 p (in Russian)
2. Ohme W (1968) Beitrag zur nutzung des elektrohudraulischen effektes in der ausbereitungstechnik. Freiberger Forschunghelfte-Leipzig, A425, pp 38–80
3. Garlei-Macayly KW(1968 September) Electrohydraulic crushing. Chem Process Eng 49:77–81
4. Vajov VF, Gafarov RR, Datskevich SY et al (2010) Electric-pulse breakdown and the breakage of granite. Tech Phys 55(6):833–838
5. Hitchov JW, Garlei-Macayly KW (1978 Juni) Effect of electrohydraulic shocks an individual solid particles. Metall Trans Sect 81 (Bulletin no 787). Harwell Berkshire Institute on Mining 8
6. Nanazashvili IKh (1990) Building materials, products and constructs: reference book. Publishing House Vysshaya Shkola, Moscow, 495 p (in Russian)
7. Vajov VF, Muratov VM, Levchenko BS et al (2012) Rock breakage by pulsed electric Discharges. J Min Sci 48(2):308–313
8. Kuretz VI, Usov AF, Zuckerman VA (2002) Electropulse disintegration of materials. Publishing House of the Kola Scientific Center of the Russian Academy of Sciences, Apatity, 324 p (in Russian)

Chapter 10
Breaking and Cutting of Rocks and Concretes

Cutting of rocks and concretes is used for dimensional processing of products or production of blocks from these materials. The ED cutting technology can effectively be used in such processes as:

- Cutting and processing of natural stones and artificial materials,
- Crack penetration and direct cutting of blocks of rock from massifs without explosive breaking,
- Stripping of rock bases under structures during building,
- Crack penetration for laying cables for different purposes.

(For example, Dresden Technological University purchased TD installation for cutting slabs of complex configurations from natural stone to pave roads in cities).

The ED breaking of rocky massif is used for the following purposes:

- Stripping of surfaces during construction of tunnels, mine trunks, and foundation pits after explosive works,
- Removal of the upper layers during repair of roads and cleaning of surfaces contaminated by chemical or radioactive wastes,
- Destruction of the sub-standard or damaged concretes and reinforced concrete products, and so on.

These directions of EDT application are at their initial stages, and the materials presented below are aimed only to demonstrate the EDT possibilities in these fields.

10.1 Breaking of Rocks and Concretes

In experiments on breaking, the IVG with stored energy up to 43.8 kJ and nominal voltage of 700 kV regulated by electrode spacing and properties of solid materials was used as a source of high-voltage impulses. The voltage impulses had the rise time (front) 0.2–0.5 μs and the time of voltage half-decay less than or equal to 50 μs.

V. Y. Ushakov et al., *Electro-discharge Technology for Drilling Wells and Concrete Destruction*, https://doi.org/10.1007/978-3-030-04591-3_10

Breakdown was initiated on the impulse front. The electrode spacing S changed from 100 to 300 mm. The electrodes represented segments of a high-voltage cable with polyethylene insulation 36 mm in diameter and copper core 6 mm in diameter. Samples were prepared from granite or brand 300 concrete.

Two variants were used for breaking:

Electrodes were placed directly on the destructible sample surface (the superimposed electrodes),

Electrodes were placed into preliminary prepared blast holes 40 mm in diameter, Fig. 10.1.

In the first variant samples with sizes 450 × 400 × 300 mm were used, and in the second variant, samples with sizes 900 × 900 × 450 mm were used. These samples were set in concrete in steel baths with sizes 1100 × 1000 × 600 mm to reduce the probability of sample fragmentation without forming slabbing craters. For the electrodes placed on the sample, the entire surface was beneath the water layer whose thickness was not less than the electrode spacing. When the electrodes were inserted deep into the blast hole, only they were filled with water. The samples broken up in the course of experiments were not taken into account.

During experiments, the following parameters were measured: the breakdown voltage U_{br}, the time before breakdown t_{br}, the period of discharge current oscillation T_{ch}, the slabbing crater volume V_{Σ}, and the number of applied impulses n. In addition, the energy W stored in the IVG, the total energy W_{Σ}, the destruction productivity Q, the total energy W_s, and the stored energy W_{spec} were calculated.

The maximal destruction efficiency was obtained for application of several impulses without changing the position of the electrodes [1, 2].

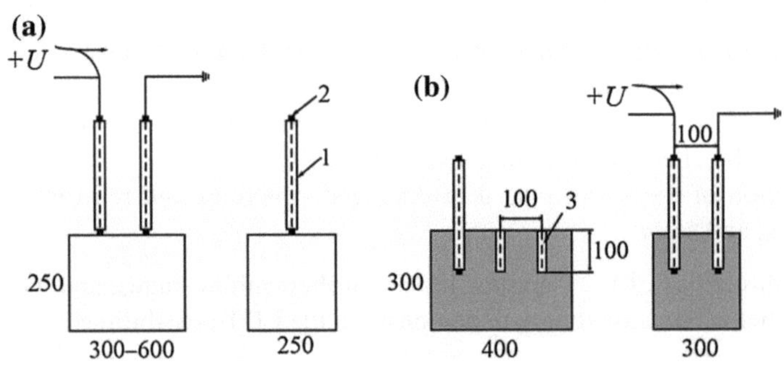

Fig. 10.1 Samples with superimposed (**a**) and inserted (**b**) electrodes

10.1.1 Breaking with Superimposed Electrodes

Figure 10.2 shows the dependences of the destruction productivity for granite (curves *1* and *2*) and concrete (curve *3*) on the distance between the electrodes at the indicated energy W_s deposited into the discharge channel.

It can be seen that a three-fold increase in S for granite at $W_s = 75$–120 J/mm (curve *1*) led to the increase of the productivity by a factor of 4.2, and at $W_s = 35$–60 J/mm (curve *2*) the productivity increased by a factor of 4.5. For concrete, Q increased by a factor of 3.76 when S doubled at $W_s = 16$–26 J/mm (curve *3*). An increase in the energy deposition contributes significantly to the increase of the destruction productivity. For example, doubling of the energy deposition for granite (curves *1*, *2*) at $S = 300$ mm caused the increase of the destruction productivity by a factor of 1.6, whereas the energy consumption increased only by a factor of 1.24. This means that the increase in the energy stored in the IVG increases the destruction volume that under considered experimental conditions reached 260 cm^3 per impulse.

Figure 10.2 shows the dependences on the interelectrode distance of the amount of energy consumed for destruction of unit volume (specific energy consumption W_s) for all experimental conditions (curves *4*–*6*). No dependence of the specific energy consumption on the distance between the electrodes was revealed. However, this result should be considered carefully because of a wide spread of data from experiment to experiment and a narrow range of change of the distance between the electrodes (100–300 mm). The dependence of the productivity of each discharge on the distance between the electrodes is more clearly pronounced. Hence, to increase the

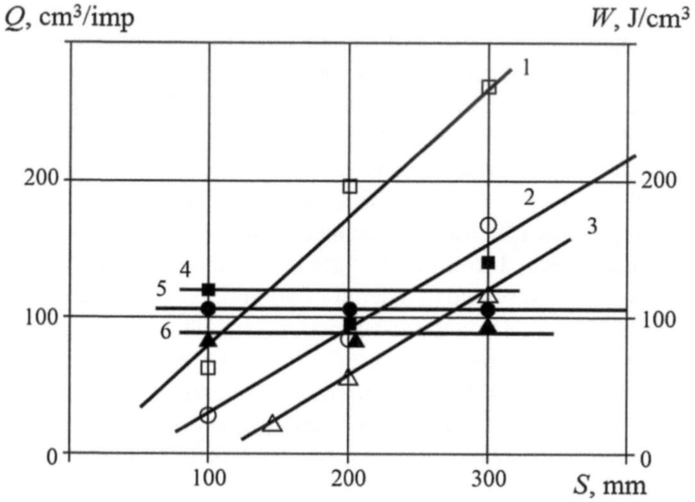

Fig. 10.2 Dependences of the destruction productivity (curves *1*–*3*) and energy consumption (curves *4*–*6*) on the distance between the electrodes for granite at $W_s = 75$–120 (curves *1* and *4*) and 35–60 J/mm (curves *2* and *5*) and for concrete at $W_s = 16$–26 J/mm (curves *3* and *6*)

Table 10.1 Energy consumption for the indicated methods of destruction of solids

Destruction method	Impact	Rotational	Explosive	Electro-hydraulic	By fire	ED
Energy consumption, J/cm^3	200–650	400–950	200–400	400–500	1500	80–138

destruction productivity or to decrease the energy consumption for a preset distance between the electrodes, it is necessary to optimize the IVG energy characteristics by the chosen criterion.

Even without optimization, the ED breaking technology consumed the least amount of energy in comparison with other technologies, Table 10.1 [3].

10.1.2 Breaking Using Blast Holes

The application of blast holes for breaking of rocks and artificial materials allows the destruction depth to be regulated by changing the depth of the blast holes. It is difficult to choose a priori the depth of the blast holes for preset distances between the electrodes and IVG energy characteristics; therefore, preliminary experiments were preformed for granite to estimate the range of depths of the blast holes to form slabbing craters of a preset depth.

In the first series of experiments, the distance between the electrodes $S = 100$ mm, the depth of the blast hole $H = 150$ mm, the deposited energy $W_s = 438$ J/mm, and the discharge energy was 43.8 kJ. After application of 11 impulses, the sample was broken along the blast hole axis without formation of slabbing craters. This means that during energy liberation in the discharge channel (the time of discharge current passage through the channel was 4.5–9.8 μs), cracks in the required number had no time to reach the sample surface, and no slabbing crater was formed.

The second series of experiments was carried out in the following conditions: the maximal depth was 150 mm, the distance between the electrodes was 300 mm, the deposited energy was 146 J/mm, and the energy stored in the IVG was 43.8 kJ. The electrodes were first placed at a depth of 50 mm. After the application of three impulses, the slabbing crater with the volume $V_\Sigma = 4186$ cm^3 (1395 cm^3/impulses) was formed, and the consumed energy was 31 J/cm^3. Then the electrodes were placed at a depth of 100 mm. After the application of one impulse, the destruction volume was 2984 cm^3/impulses, and the consumed energy was 14.7 J/cm^3. The destruction volume for the electrodes placed at a maximal depth of 150 mm after the application of nine impulses were 1791 cm^3 or 199 cm^3/impulses, and the consumed energy was 292 J/cm^3. Hence, the blast holes 50–100 mm deep can be considered optimal for breaking using the employed discharge. All subsequent experiments were performed with blast hole depths of 50 and 100 mm and the distances between the electrodes of 200 and 300 mm.

Table 10.2 presents results of breaking of granite and concrete. The number of applied impulses changed from 6 to 9 depending on the sample states during tests, and the number of samples in the batch was 3–6.

From Table 10.2 it follows that for granite with $S = 300$ mm and approximately the same deposited energy, the destruction productivity increased by a factor of 2.4, and the energy consumption decreased by a factor of 3.2 when the blast hole depth was doubled. For $S = 200$ mm and $H_{bh.} = 100$ mm, the destruction characteristics were worse even when the energy deposition increased by a factor of 1.75 for $S =$ 300 mm: the productivity was by a factor of 1.22 smaller and the energy consumption was by a factor of 1.53 larger. Hence, the increase of the electrode spacing with optimization of the discharge energy characteristics is an attractive method of increasing the efficiency of breaking of rock with blast holes.

It is of interest to consider the influence of the tilt angle of blast holes to the sample surface on the efficiency of forming of slabbing craters. The blast holes 40 mm in diameter were drilled in granite samples at an angle of $40 \pm 5°$ to the sample surface to a depth of 50 mm. The electrodes were spaced at 300 mm. From Table 10.2 it follows that the productivity of destruction with inclined blast holes increased by a factor of 1.28 for the same deposited energy, whereas the energy consumption decreased by a factor of 1.71. It seems likely that in this case, the preliminary prepared tilted blast hole that allows the destruction area and volume to be increased thereby increasing the productivity of breaking influences greatly.

To reveal the influence of physicomechanical characteristics of solids on the breaking process, samples from brand 300 concrete were tested. Their specific mechanical strength under uniaxial compression was by a factor of 2.74 less than of granite used in the above-described experiments. From Table 10.2 it follows that for energy deposition by 24% less than for granite, the productivity of breaking is doubled compared to that of concrete, and the energy consumption was by a factor of 2.89 smaller.

Thus, the use of blast holes increases the productivity of breaking many times and decreases the energy consumed for ED destruction. (Without taking into account the energy consumed for preparation of blast holes.). The destruction process should be optimized for the characteristic recognized as a determining one.

Table 10.2 Results of breaking of granite and concrete

$H_{b.h.}$, mm	S, mm	U_{br}, kV	W, kJ	W_{spec}, J/cm^3	Q, cm^3/impulses	W_s, J/mm
Granite: vertical blast holes						
50	300	630	35.4	105.2	416	118
100	300	600	33.9	33.8	1005	112
100	200	610	39.5	51.8	820.7	197
Granite: blast holes at an angle of 40±5°						
50	300	600	32.7	61.3	534	109
Concrete: vertical blast holes						
100	200	460	30.0	17.9	1666	150

10.2 Cutting

10.2.1 *Cutting with a Cyclically Movable Electrode System*

Despite considerable progress in technique and technology of penetration into cracks and tranches in rocks and concretes, the productivity and cost of these works remain unsatisfactory. The main reasons here are fast deterioration of the working tool and its low efficiency. The characteristics of the traditional mechanical methods of cutting of rocks are presented in Table 10.3 (for sandstone), where W_s is the energy consumed per unit of the newly formed surface, and P is the productivity of forming of the new surface.

The ED method of cutting of nonconductive solids can appear competitive among the new methods (based on the new principles) that are being searched and investigated continuously.

In [4] eight possible types of devices for cutting and dimensional surface processing were listed and briefly considered.

However, only three cutter designs were tested, including lamellar and rod cutters and their modifications. Cutting was carried out in transformer or diesel oil. In the first works on ED cutting, the cutters with electrodes whose length was equal to the length of the crack cut for one cycle was used. After that the working tool (cutter) moved to a new position to continue cutting (To designate this regime, the term "cyclic" was used in the monograph.).

Since the unsuitability of these liquids for the working medium in the conditions of production was quite clear, subsequent investigations and developments were focused on the application of technical water as a working liquid. In this case, two serious problems arose, namely, large losses of energy because of low ohmic resistance of the cutter for large non-insulated surfaces and breaking of cutter insulation near the active zone in the attempt to reduce the area of non-insulated surface. The best characteristics in this aspect have cutters with rod electrodes (rod cutters). In cutting with a rod cutter in water with specific resistance $\rho = (0.6–0.9) \times 10^4$ Ω cm of the crack 0.35 m long and 50 mm wide, the specific energy consumption was 4–6 kWh/m^2 for sandstone and 3.5–4.5 kWh/m^2 for limestone. However, with such cutter design, it was difficult to remove the crushed rock from the crack, and during cutting, the cutter often "hanged" on the crack walls. The increase of the cutter length and hence, the length of the crack cut for one cycle was limited by the ohmic resistance of the

Table 10.3 Characteristics of mechanical methods of cutting

Method of cutting	W_s, kW/m^2	P, m^2/h	Crack width, mm
Cable saw	4.6	3.0	8–11
Pendulum saw	5.2	2.4	6–10
Chain saw	2.6	10	<42
Disc saw	4.4	1.0	3–5

cutter. For long cutters with the use of technical water with low ohmic resistance, losses of energy stored in the IVG became inadmissibly large because of conductivity currents. In first studies on ED cutting, the impulse energy was no less than 10^2–10^3 J. This necessitated the application of bulky IVG unacceptable for mobile installations. To develop an efficient installation for ED cutting with the use of water, the number of cutting electrodes and their area must be decreased, thereby reducing the energy to be stored in the IVG. The minimal number of electrodes is two, but in this case, the length of the cut, approximately equal to the electrode spacing, is very small, since it is limited by the working voltage of the IVG and the rock breakdown voltage in the interelectrode gap. Since no acceptable solutions to this problem had been found, interest to this field of EDT application had been lost for many years.

10.2.2 Cutting with a Continuously Movable Electrode System

The idea of ED cutting was partially rehabilitated when it was suggested to move continuously the two-electrode cutter along the cutting trajectory in the process of cutting. In this case, the electrodes are placed along the axis of the crack being cut, thereby excluding hanging of the electrode system on the roughnesses of crack walls and eliminating the problem of reliable operation of solid insulation [5].

Experiments were performed using the test stand comprising an IVG, a vessel with water into which the examined sample of solid was placed, a system of movable electrodes, and washing, measuring, and protecting systems. The cutting tool was a two-electrode rod system, Fig. 10.3.

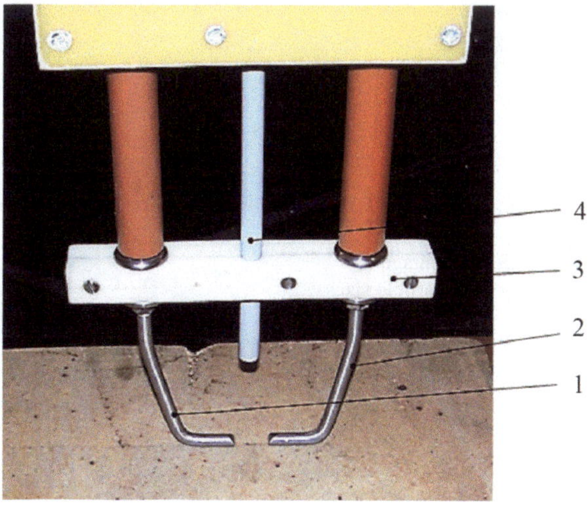

Fig. 10.3 Electrode system comprising electrodes from stainless steel 1, insulation holders 2 and 3, and washing tube 4

The electrode shapes were chosen to minimize their snag on crack walls and bottom. The electrode spacing could be changed within 20–28 mm. The electrode system moved with adjustable velocity along the surface of the solid being processed (granite, sandstone, gabbro, or brand 300 concrete) by means of a drive. Technical (tap) water with specific resistance (6.5–30) × 10^3 Ω cm was used as a working liquid. The photograph of the sandstone sample with two cracks cut by the ED method is shown in Fig. 10.4.

The crack walls are rough. The degree of roughness depends primarily on the electrode spacing and the energy in the discharge. For the impulse energy $W =$ 30–125 J and the gap length up to 28 mm, the maximal depth of protrusions and recesses on the crack walls was 2 mm for sandstone and 6 mm for granite. The estimated roughness of crack walls was refined by sieve analysis of destruction products (sludge), Fig. 10.5.

Experiments with continuously moving two-electrode system were carried out with energy stored in the IVG $W = 44.5$ J. The displacement step m was always less than S. The required number of applied voltage impulses for the electrode system

Fig. 10.4 Crack cut in sandstone

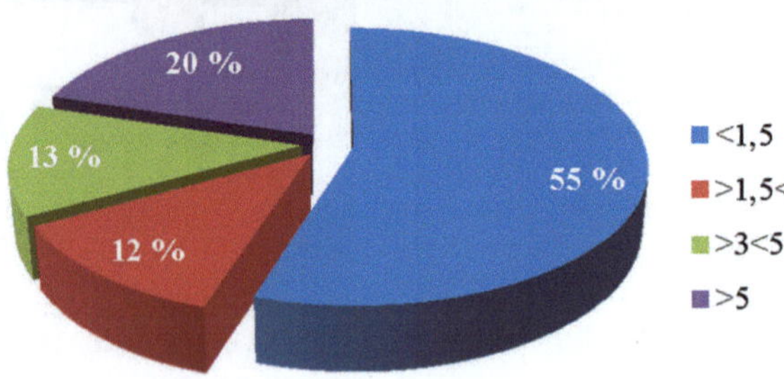

Fig. 10.5 Percentage of sludge granite particle sizes

displaced by S depended on the amount of energy W_{ieg} liberated in the interelectrode gap. The dependences of the relative average destruction depth H/S on the number of impulses for three S values are shown in Fig. 10.6 for sandstone.

It can be seen that for motionless electrodes these dependences are saturated at 7–12 impulses (curves *1* and *2* in Fig. 10.6). For movable electrode system, saturation was not observed for the examined range of n values.

In cutting of rocks by any method, it is desirable to decrease the crack width to reduce the consumption of energy and material being processed. Figure 10.7 shows the dependences of the change in the crack width on the energy stored in the IVG for the same displacement step and on the displacement step of the movable electrode system at the same energy stored in the IVG. In this case, the distance between the cutter electrodes was 28 mm for sandstone; for granite, it was 23 mm. As can be seen from the plots, the crack width decreased with increasing step of the electrode system displacement (curve *2*). This is quite explainable, since in this case the number of impulses acting on the rock in the interelectrode gap decreases; as a result, the volume of the rock chipping away from the crack walls decreases. The crack width increases with IVG energy (curve *1*), because the productivity of the single impulse increases owing to the increased length of the formed cracks and of the destruction zone around the discharge channel.

In the above-described experiments, the energy stored in the IVG changed from 28 to 125 J, which was by a factor of 10–16 less than in the experiments with cyclic movable electrodes.

It is expedient to characterize the efficiency of cutting by the energy consumed per unit area of newly formed surface ω_S. The dependences of ω_S on the energy $W_\Sigma = W{\cdot}n$ consumed for continuous displacement of the electrode system at the distance

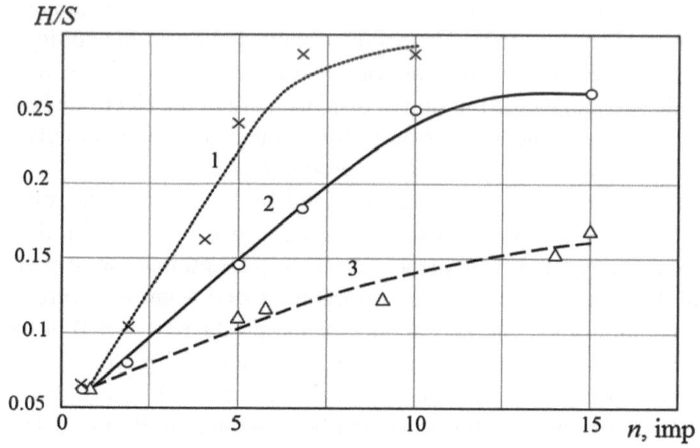

Fig. 10.6 Dependences of the change of relative destruction depth on the number of impulses at W = 44.5 J for S = 20 (curve *1*), 28 (curve *2*, motionless electrodes), and 28 mm (curve *3*, movable electrodes)

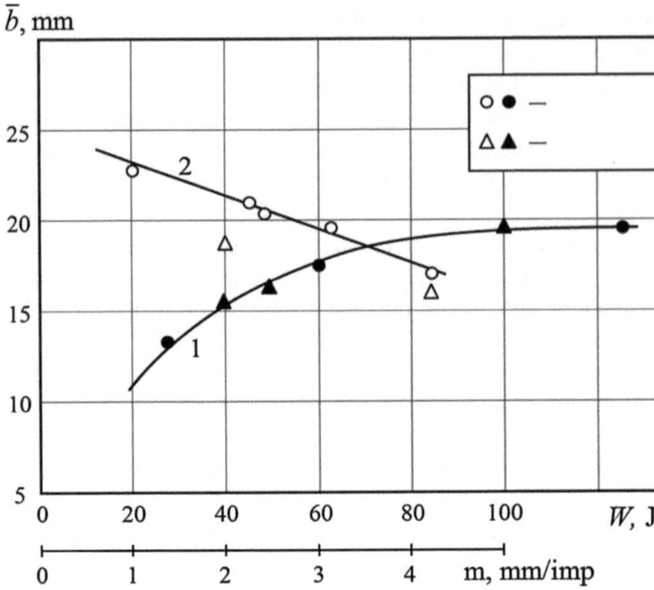

Fig. 10.7 Dependence of the crack width on the energy stored in the IVG and on the displacement step for granite (S = 23 mm) and sandstone (S = 28 mm). Here curve *1* shows $\bar{b} = f(W)$ and curve *2* shows $\bar{b} = f(m)$

equal to the interelectrode gap length are shown in Fig. 10.8. The form of the dependences testifies to the possibility of optimization of the specific energy consumption individual for each material. The specific energy consumed for destruction for both cyclic and continuously movable electrodes increased linearly with the acoustic rigidity of materials, Fig. 10.9 [62]. The results obtained allowed the cutting process to be optimized by the energy consumption for different rocks, that is, to choose the interelectrode gap length and the step of displacement of the electrode system and to formulate the requirements to the energy per impulse and to the impulse repetition frequency to provide the required cutting productivity.

The cutting productivity P is characterized by the area of the surface newly formed in unit of time. Since in the above-described experiments the IVG provided the impulse repetition frequency less than the optimal one, it was necessary to estimate P from the area of the surface produced upon exposure to a single impulse (cm^2/impulses). The dependence of the productivity on $W{\cdot}n$ had a maximum observed, for example, at $W{\cdot}n \sim 900$ J for granite, Fig. 10.10, that is, it was similar to the dependence of the energy consumption on $W{\cdot}n$, Fig. 10.7 (curve *1*). The productivity per single impulse, naturally, depended on W, Fig. 10.9.

The results obtained by different authors using the ED technology of cutting of rocks in diesel oil and water are presented in Table 10.4, where S is the electrode spacing, W is the energy stored in the IVG, W_{spec} is the specific energy consumed for cutting, and w is the energy consumed per unit area of the newly formed surface.

Table 10.4 Characteristics of different electrode systems

	Electrode system	S, mm	W, J	W_{spec}, J/cm^3	ω_S, J/cm^2	b, mm	Material	Medium	Refs.
1	Plate	5–25	200–500	2150–1660	4–6	50	Sandstone, marble, granite	Diesel oil	[4]
2	Multi-rod	20–50	800–1500	1440–590	4–6	50	Sandstone, granite, marble, concrete	Diesel oil, water	[4]
3	Two-rod movable	20–28	44.5–50	250–350	0.75–2.6	15–20	Sandstone, granite, gabbro	Water	[6]

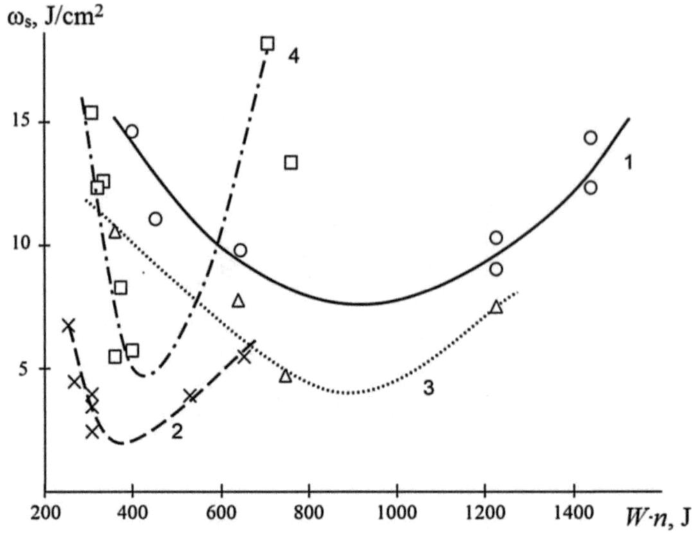

Fig. 10.8 Dependence of the specific energy consumption (for unit newly formed surface) on the energy consumed for cutting of the crack whose length is equal to the interelectrode gap length. Here curve *1* is for granite, curve *2* is for sandstone, curve *3* is for gabbro, and curve *4* is for concrete

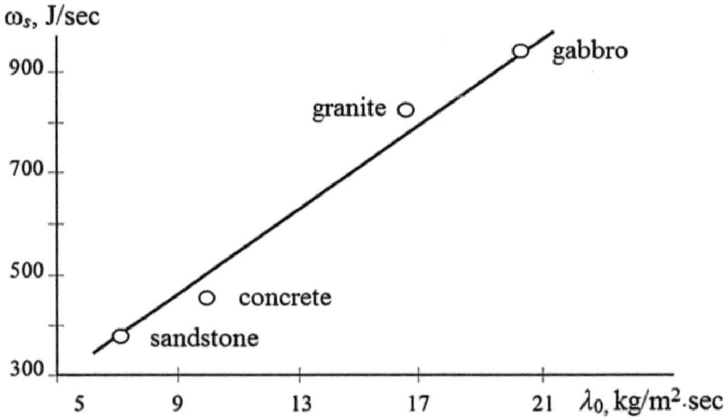

Fig. 10.9 Dependence of the specific energy consumption on the acoustic rigidity of materials

From Table 10.4 it follows that the specific energy consumption w for the same discharge gap lengths and the two-rod continuously movable electrode system is by a factor of 2.3–8.0 less than for the multi-electrode cyclically movable system. Moreover, the energy in the discharge for the two-rod electrodes decreased by 18–30 times. The best cutting characteristics of the two-electrode system were obtained due to (a) the decrease of the crack width by a factor of 2.5–2.7, (b) the absence of "hanging" of the electrode system on the crack walls, (c) the increase in the

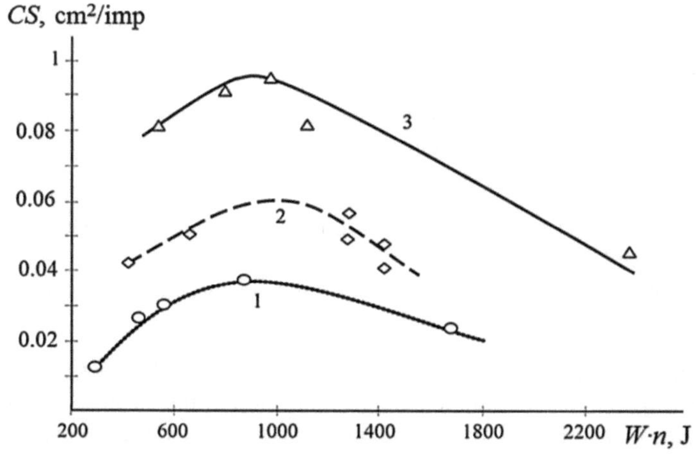

Fig. 10.10 Dependence of the productivity of granite cutting with electrodes displaced by the interelectrode gap length on the consumed energy for W = 54 (curve *1*), 72 (curve *2*), and 125 J (curve *3*)

probability of discharge channel penetration into the rock, and (d) the decrease of the energy losses caused by the conductivity currents for cutting in water.

The results stated above demonstrate that optimization of the cutting process is the problem whose solution depends on a large number of factors. The empirical approach to the determination of the displacement step m of the two-electrode system and of the energy W stored in the IVG was developed for optimal energy (W_{opt}) consumed for cutting of different rocks. The consumed energy linearly increased with the hardness ratio f_p and acoustic rigidity, Fig. 10.9 [6]. For the reference value of the optimal specific energy consumption W_{opt1}, its value for sandstone was taken. Then for other stronger rocks ($20 \geq f_p \geq 4$–5), the specific energy consumption W_{opti} was determined by the formula

$$W_{opti}/W_{opt1} = 0.38 + 0.13 f_p, \tag{10.1}$$

and the energy stored in the IVG was determined by the formula

$$W = \left(0.38 + 0.13 f_p\right) W_{opt1}/n, \tag{10.2}$$

where $n = S/m$ is the number of impulses necessary for the electrode system to travel the interelectrode distance S which depends on the displacement step. The displacement step of the electrode is calculated as follows:

$$m = (0.33 - 1.11) W \cdot S / W_{opt1}, \tag{10.3}$$

where the proportionality coefficient (0.33–1.11) depends on the rock hardness: for the strongest rock, for example, for microquartzite, it is equal to 0.33, and it is 1.11 for the least strong rock, for example, sandstone.

The results of experimental cutting by cutters of different designs under various operating conditions have allowed us to formulate the main advantages of cutting by two-electrode continuously movable rod cutters as follows:

(1) Much higher energy efficiency of cutting and smaller energy stored in the IVG (10 folds and more),
(2) Less capital expenditures required for the development of installations and improved reliability of their operation.

Two important specific features were established: (a) the energy consumption and the cutting rate depend on the displacement step of the electrode system whose optimal value is determined by the discharge gap length and the energy stored in the IVG, and (b) the productivity and the cutting rate with a preset displacement step size are determined by the impulse repetition frequency and the corresponding velocity of motion of the electrode system.

Based on the results of experimental ED cutting, the idea of motion of the working tool in the horizontal plane in the process of well drilling was suggested. In this case, the drilling implement or only the drilling head are rotated with velocity changing within the limits determined by the impulse repetition frequency and the displacement step of the drilling head. The head itself has a small number of the electrodes, which is very attractive for drilling with water. Preliminary experiments demonstrated that the promise of such processing method consists in the following:

- Use of water as a working liquid,
- Decrease of the number of electrodes of the drilling head that reduces the energy losses in the pre-breakdown stage and the distortion of the acting impulse shape,
- Absence of "hanging" of the drilling implement on roughnesses of the well bottom and walls,
- Increase in the efficiency (productivity) of rock destruction,
- Considerable decrease of the energy stored in the IVG and, as a consequence, the decreased cost of its construction.

References

1. Jgun DV, Jurkov MY, Lopatin VV (2002) Application of pulsed discharges for materials cutting. In: European pulsed power symposium. French German Research Institute, Saint Louis, France, pp 22/1–22/4
2. Zhurkov MY, Vazhov VF, Kozachenko SN (2002) Investigation of the stone cutting conditions with the electrical pulse discharges of low energy. In: VIII International conference modern technique and technologies (MTT 2002), Tomsk, pp 27–28
3. Siomkin BV, Usov AF, Kuretz VI (1995) Bases of electropulse destruction of materials. St. Petersburg, Nauka, 276 p (in Russian)

4. Kuretz VI, Usov AF, Zuckerman VA (2002) Electropulse disintegration of materials. Publishing House of the Kola Scientific Center of the Russian Academy of Sciences, Apatity, 324 p. (in Russian)
5. Vazhov VF, Zhurkov MY, Lopatin VV et al (2008) Electric-discharge cutting of rocks. J Min Sci 44(2):176–182
6. Vajov VF, Zinoviev NT, Ushakov VY (2016) In: Ushakov VY (ed) Electrodischarge technology for drilling wells and fracture of reinforced concrete products: monograph. Publishing House of Tomsk Polytechnic University, 312 p (in Russian)

Conclusion

Results of investigations and laboratory and field studies (outdoor tests of the EDT) have demonstrated the promising character of the EDT application for drilling wells, destruction of reinforced concrete products, and breaking, and cutting. Stages of its development for solving these four problems differ, but no one of them has principal limitations for further increase of the efficiency, reliability, lifetime, and convenience and safety of operation. The optimism of such estimations is based on a great progress reached since the beginning of works on the EDT up to now in technical areas on which the success in overcoming still existing obstacles for wide application of this technology depends. The matter in point is primarily perfection of high-voltage impulse technologies, development of structural and insulation materials with improved properties, and increase in the flexibility of the automated systems of control over the technological processes.

V. Y Ushakov et al., *Electro-discharge Technology for Drilling Wells and Concrete Destruction*, https://doi.org/10.1007/978-3-030-04591-3

Appendix A
Photographs of Installations for RCP Destruction

See Figs. A.1, A.2, A.3, A.4, A.5, A.6 and A.7.

Fig. A.1 Installation for concrete destruction (Test site of HVRI, Tomsk, Russia, 1986)

V. Y Ushakov et al., *Electro-discharge Technology for Drilling Wells and Concrete Destruction*, https://doi.org/10.1007/978-3-030-04591-3

Fig. A.2 Test site of the HVRI, Tomsk, Russia, 1990

Fig. A.3 Leninakan, Armenia, 1988

Fig. A.4 Kobe, Japan, 1995

Fig. A.5 Unterlüß, Germany, 1997

Fig. A.6 Seversk, Russia, 2011

Fig. A.7 Tselinograd, Kazakhstan, 1992

Appendix B
Photographs of Some RCPs in Different Destruction Stages

See Figs. B.1, B.2, B.3, B.4, B.5, B.6, B.7, B.8 and B.9.

Fig. B.1 Contact network support

V. Y Ushakov et al., *Electro-discharge Technology for Drilling Wells and Concrete Destruction*, https://doi.org/10.1007/978-3-030-04591-3

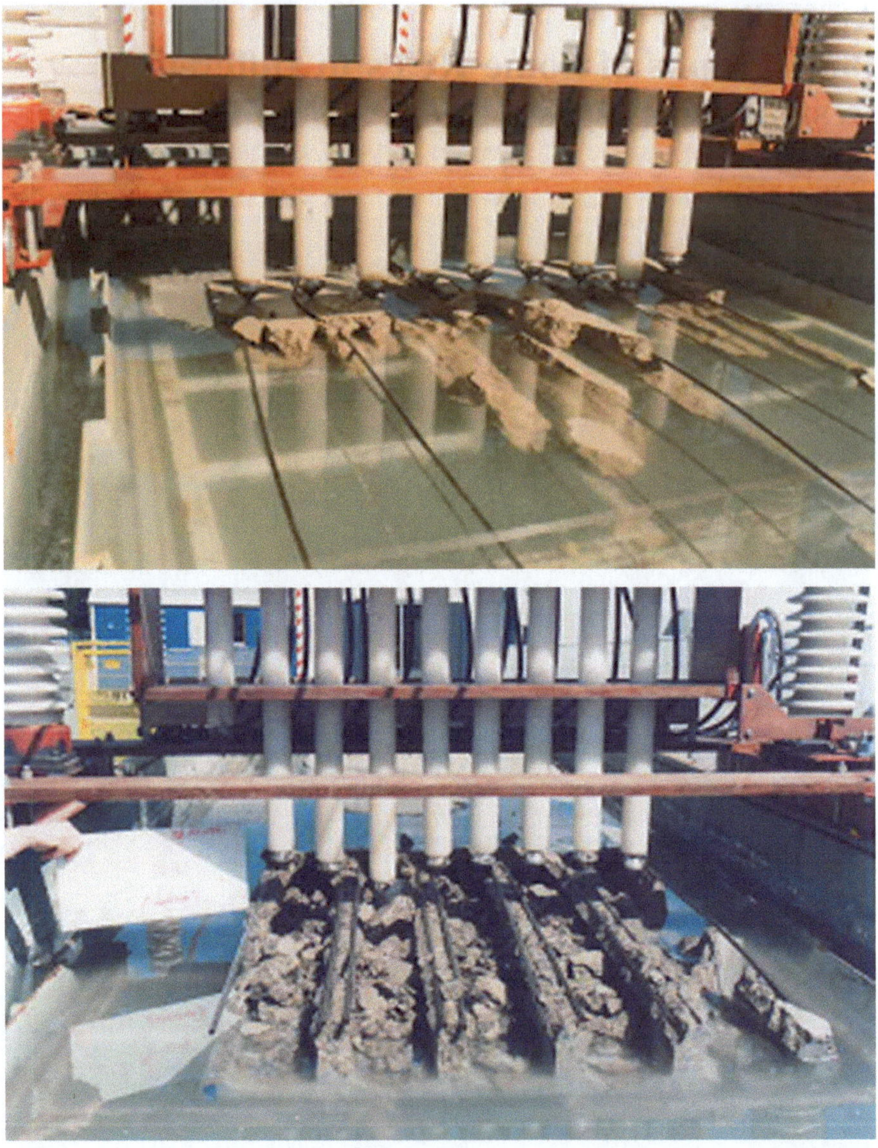

Fig. B.2 Floor slab (Unterlüβ, Germany)

Fig. B.3 Floor slab (Tomsk, Russia, 1986)

Fig. B.4 Single-layer reinforced floor slab (Tomsk, Russia, 1986)

Fig. B.5 Lightning network support (fragment): **a** before destruction and **b** after destruction

Fig. B.6 Three fragments of the contact network support, one of which was destructed (Tomsk, Russia, 2012)

Fig. B.7 Contact network fragments (Tomsk, Russia, 2012)

Fig. B.8 Utilization products

Fig. B.9 Embedded parts extracted from RCPs

Bibliography

1. Smolyanytsky AA, Genkin EE (1974) Investigation of the nature of the destruction of rocks during an electric explosion. Gorny J Izvest vuzov (No. 9):69–71 (in Russian)
2. Smirnov SM, Terentyev PV (1964) High voltage pulse generators. Publishing House Energiya, Moscow, 239 p. (in Russian)

V. Y Ushakov et al., *Electro-discharge Technology for Drilling Wells and Concrete Destruction*, https://doi.org/10.1007/978-3-030-04591-3

MIX
Papier aus verantwortungsvollen Quellen
Paper from responsible sources
FSC® C105338

If you have any concerns about our products,
you can contact us on
ProductSafety@springernature.com

In case Publisher is established outside the EU,
the EU authorized representative is:
Springer Nature Customer Service Center GmbH
Europaplatz 3, 69115 Heidelberg, Germany

Printed by Libri Plureos GmbH
in Hamburg, Germany